S0-AGF-864

MONOGRAPHS ON CRYOGENICS
in collaboration with
the British Cryogenics Council

Editor: R. G. Scurlock

University of Southampton

Ford (USA) design of maglev vehicle

Warwick test track

Model of Canadian maglev vehicle

Advanced Passenger Train (B.R.)

High-speed ground transport

MAGNETIC LEVITATION FOR RAIL TRANSPORT

R. G. RHODES
University of Warwick
AND
B. E. MULHALL
University of Surrey

CLARENDON PRESS OXFORD · 1981

Oxford University Press, Walton Street, Oxford OX2 6DP

London Glasgow New York Toronto
Delhi Bombay Calcutta Madras Karachi
Kuala Lumpur Singapore Hong Kong Tokyo
Nairobi Dar es Salaam Cape Town Salisbury
Melbourne Auckland

and associate companies in
Beirut Berlin Ibadan Mexico City

© R. G. Rhodes and B. E. Mulhall 1981

Published in the United States
by Oxford University Press, New York

All rights reserved. No part of this publication may be
reproduced, stored in a retrieval system, or transmitted, in any
form or by any means, electronic, mechanical, photocopying,
recording, or otherwise, without the prior permission of Oxford
University Press

British Library Cataloguing in Publication Data

Rhodes, R. G.
   1. Magnetic levitation vehicles   2. Locomotives
   I. Title   II. Mulhall, B. E.
   625.2'6   TJ609

ISBN 0-19-854802-8

Typeset by Cotswold Typesetting Ltd, Gloucester and
Printed in Great Britain
by Eric Buckley,
Printer to the University,
Oxford University Press, Oxford

CONTENTS

Preface   vii

1   High-speed land transport   1
   1.1   Transport in the future   1
   1.2   Advanced ground transport   3
   1.3   Wheel-on-rail systems   4
   1.4   Levitated systems   5

2   Introduction to magnetic levitation   8
   2.1   Types of magnetic levitation   8
   2.2   Permanent magnets   9
   2.3   Electromagnetic attraction   10
   2.4   A.C. (repulsion) systems   14
   2.5   Electrodynamic (d.c.) repulsion   16
   2.6   Mixed-μ suspension   21

3   Electrodynamic levitation   24
   3.1   Theory   24
   3.2   Null-flux systems   29
   3.3   Experimental verification   32
   3.4   Scaling problems with model systems   36

4   Vehicle and guideway design   37
   4.1   Types of guideway   37
   4.2   Estimation of force characteristics   41
   4.3   Guidance forces   43
   4.4   Vehicle stability and aerodynamic forces   46
   4.5   Ride quality   48
   4.6   Superconducting magnets   52
   4.7   Magnetic screening   55

5   Propulsion   58
   5.1   Power requirements   58
   5.2   Powered vehicles   59
   5.3   Passive vehicles   61
   5.4   Linear synchronous machines   62
   5.5   Miscellaneous propulsion   66

6   Maglev present and future   68
   6.1   Introduction   68
   6.2   The USA   68
   6.3   Japan   69
   6.4   Canada   75

6.5 Western Germany	83
6.6 The United Kingdom	90
7 Future prospects	94
Bibliography	97
Index	101

PREFACE

Together with the provision of food and shelter, a major part of man's economic activity is devoted to transport, moving people and goods from one place to another, and the supporting industry which makes this movement possible. It can also be argued that many of the threats to our present social and economic organization are attributable to the same transport industries. It is not surprising, then, that any idea which promises to improve the means of movement of people or goods should attract wide interest. In this monograph we are concerned with a new form of guided land transport, although it must be recognized that the costs of introducing novel 'railway' systems offering significant benefits in speed and efficiency are very high indeed. It is perhaps not surprising that the most significant efforts to develop a levitated railway are being made in Japan and West Germany, followed somewhat behind by Canada, the USA and the UK.

The first practical method investigated of overcoming the limitations of wheel-on-rail traction was the use of air-cushions to support a vehicle above a guideway. Although air-cushion levitation has been very successfully exploited in marine hovercraft, it has been a noticeable failure for guided land transport. The possibility of supporting a vehicle by magnetic forces had been postulated from the early part of this century, but it has only received serious consideration in the past decade or so; this is basically a result of the development of high-field superconductors and the associated cryogenic technology. There is little doubt that the superconducting magnet is vital for the development of the system of electrodynamic magnetic levitation and linear motor propulsion which we discuss.

The present status of magnetic levitation is that not only has a great deal of basic study and experimentation been carried out by research groups in several countries but also some quite large scale developments have been undertaken, as described in the text. However, in view of the enormous costs involved in a full-scale system, further progress appears now to be more dependent on political and economic considerations than on the technologies involved. It thus seems timely to review this field of advanced transport and to describe and compare some of the different designs that have been put forward.

As the number of research groups involved in magnetic levitation is relatively small it would perhaps be inappropriate and presumptuous on our part to present a discussion, in great depth, of the various proposals. Therefore, rather than to go into considerable detail of the many aspects of a rather broad subject, our principal aim has been to give a wider ranging presentation for the more general reader. Taking this standpoint, no attempt has been made to compile an exhaustive bibliography; instead, a selection of references has been included which should lead the interested reader fairly rapidly to an appreciation of the extensive literature on the subject. It is to be hoped, also, that our own position, being associated for the past decade with a relatively small research group in the UK, has enabled us to describe reasonably accurately and equitably the efforts of all those elsewhere who have contributed so much to the subject and thus helped to make this book possible.

Although the views presented here are our own, and we alone are responsible for any omissions or errors, we would, nevertheless, like to acknowledge with gratitude the very considerable help and support received from many sources worldwide. We would particularly mention Ed Abel, Jeff Howell, Joe Mahtani, Henry Woodgate, Don Newman and Jan Rakels among the present and former members of the Warwick Maglev team. For valuable discussions, and the provision of information, reports, and illustrations on their respective research and development programmes we would like to thank the following: the Canadian Institute of Guided Ground Transport and the National Research Council (NRC) of Canada and individual members of the team, including D. L. Atherton and A. R. Eastham of Queen's University, G. L. Slemon of Toronto, B-T. Ooi of McGill, and W. F. Hayes (NRC, Ottawa); the Japanese group, including Y. Kyotani, of Japanese National Railways, K. Oshima, of Tokyo University and F. Irie of Kyushu University; the German group based at the Siemens Laboratories including C. Albrecht, C. P. Parsch and L. Urankar; members of the several American groups, including J. R. Powell, J. R. Reitz and R. H. Borcherts among many others. We would also specifically thank Macdonald Futura Publishers, Ltd., for permission to reproduce Fig. 2.1, J. R. Powell (Brookhaven National Laboratory) and Plenum Publishing Corporation for permission to reproduce Fig. 2.6, M. N. Wilson (Rutherford and Appleton Laboratories) and M. K. Bevir (Culham Laboratory) for providing Fig. 2.7, J. R. Reitz (Ford Motor Co.) and the American Institute of Physics for permission to print Table 3.1, and J. R. Reitz for the photograph of the US design shown in the Frontispiece, C. Albrecht and the Magnetic Levitation Project Group of Siemens-BBC-AEG for permission to use Figs. 3.5, 3.6, 6.9, 6.10, 6.11, and 6.12 and Tables 6.3, 6.4, and 6.5, and Springer-Verlag for Figs. 3.5 and 3.6. It is also a pleasure to thank the secretarial staff at Surrey and Warwick for their patience and co-operation in producing and editing the typescript, the Warwick Drawing Office staff, and the Staff of the Oxford University Press for their help in producing the book.

Finally, we would like to express our sincere gratitude to the Wolfson Foundation for the initial grant which made possible the start of the maglev project at Warwick, and we would also like to acknowledge the Science Research Council for their subsequent grants which enabled the research to continue.

R. G. Rhodes
B. E. Mulhall
April 1981.

1

HIGH-SPEED LAND TRANSPORT

1.1 Transport in the future

Transport is one of the foundations of our present-day society, influencing greatly both social and economic conditions. However, the problems of moving ever-increasing numbers of people from place to place, rapidly and efficiently, are now becoming so acute as to demand increasing attention in all the developed countries of the world. Congestion and delays on the roads, in the airways and at airports, at railway stations, etc., lead not only to frustrations and inefficiency, but also to an increased risk of accident, so that, in spite of measures to improve existing ways of travelling, the social and environmental problems directly linked with modern transport continue to worsen. In particular, there is a widely held opinion, shared by the authors, that our growing reliance on cars and air travel for medium-length journeys is producing an accelerating deterioration in our social habits and life style.

The situation is being further aggravated by the predicted shortage of oil accompanied by its escalating cost. This implies that oil-fuelled vehicles, such as cars and aeroplanes, will not by themselves be able to meet people's demand for transport by the start of the next century, by which time it is predicted that oil supplies will be seriously depleted. It seems highly unlikely that fresh reserves of oil can be found to meet the total demands of the future even at the present rate of consumption. Whatever the truth, or otherwise, in detail of the energy crisis, it is evident that supplies of cheap and plentiful oil can no longer be taken for granted and that, on present-day forecasting, it will become a very precious and costly commodity by the next century, if not before. If our concern for reducing our dependence on oil is at all serious, then transport, accounting as it does for around 25 per cent of *all* energy consumption, must be one of the prime targets. The present profligate consumption of so much oil in our conventional transport modes must be drastically reduced if the natural reserves are not to disappear entirely within a generation.

Faced with these considerations the thinking in a number of the more advanced countries has turned to completely new, ground-based, systems of mass transport. Preferably these should not rely entirely on one particular source of energy; in other words, they should be electrically driven, as the generation of electric power is not tied to any one source of fuel. At the same time they must be able to meet the more stringent requirements of not polluting the environment, as well as equalling or improving on the safety record of present-day railways, which are statistically one of the safest forms of travel. Finally, they should take account of the fact that, as society continues to change, so its demand for travel and transport will continue to expand. One way of meeting some of these demands in the future, within the limitations of available land and other resources, is a mass transport system capable of speeds well in excess of those reached by conventional railway trains. High speed is desirable for economic reasons, and this is particularly so for today's industrial societies; it also has considerable passenger appeal for its own sake. Throughout history, whether man has been travelling by land, sea, or air, speed has always commanded a premium, and has been a major influence in the choice of means.

As our oil supplies become depleted and fuel costs rise, dependence on the car and the aeroplane may be expected to diminish, and travel may eventually emerge into the following pattern. For relatively short journeys, say up to 100 km, the car will continue to be used, albeit possibly in a different form from today (e.g. electrically propelled) or by fewer people, primarily because of its instant availability and door-to-door convenience. Likewise, for very long distances, of say 1000 km and beyond, the jet aeroplane travelling at 1000 km/h will continue to be the preferred mode of travel, as its high speed compensates for the delays at airports and the relatively slow journey to and from city centres. This is illustrated by the graphs of Fig. 1.1. However, in the intermediate range of distances, say between 100 and 1000 km, which includes the majority of inter-city journeys in most countries or industrialized regions of the world, the natural advantages of either car or 'plane cannot be fully exploited; it is in this sector that an advanced ground transport system has the

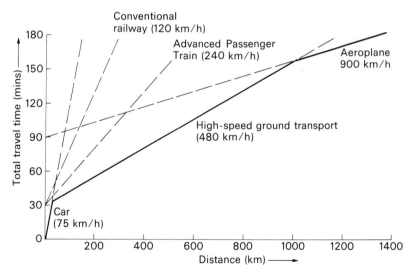

Fig. 1.1   Total journey time against distance for various modes of transport

potential to out-perform both, not only in terms of travel time, but also in safety, comfort, and efficiency.

1.2   Advanced ground transport

The advanced ground transport systems being proposed all comprise a special guideway, traversed for at least the greater part of its length by its own special vehicles (though there may be compatibility with other systems at places such as termini). One of the principal advantages of a guided system is its inherent suitability for automatic control. This allows much higher speeds of operation and hence larger capacities to be obtained with safety than is possible for vehicles, perhaps of widely different characteristics, each controlled manually by drivers with varying skills and weaknesses. It is also reasonable to assume that the capital expended in providing such a double-track guideway system would be lower than that for a motorway of comparable capacity. Furthermore, in terms of resource requirements and land usage the railway is the preferred system.

Short air journeys over land would seem to have little future in competition with a high-speed ground system because of the delays in air travel caused by the location of airports well outside city limits, and by the waiting times involved in the take-off and landing of aircraft. Not only would the ground system provide city centre to city centre connections, but it would be virtually independent of weather conditions. Recent statistics on travel in Europe and America show that domestic short-haul air transport declines when in competition with the railways after only modest improvements in the speed and service of the latter. Thus there would appear to be little doubt that the introduction of very fast ground transport would pose a serious commercial threat to the short-haul airlines.

The difficulties of expanding internal air services are also becoming increasingly serious, as highlighted by the continuing controversy over a third London airport and the violent reaction against the use of Narita, the new Tokyo airport, before it was eventually opened. The present policy of building new airports or extending existing ones has clearly shown up the adverse effects of noise and pollution and the general spoiling of the environment in the neighbourhood of airports. The situation has been further aggravated by the future landing and taking-off of supersonic aircraft, so that in some countries there is fierce opposition to the use of their air space by such planes.

There is, therefore, considerable interest in advanced ground transport systems which are capable of speeds leading to actual travel times comparable with those of aircraft. Of all the modes of internal transport, the railways are best suited to exploit automation to the maximum. The consequent precision in their operation, route setting, and traffic control may all be directed towards higher transport capacity. An automatically controlled, very high-speed, wheeled railway system, with average speeds of up to 160 km/h yet capable of meeting the currently accepted standards of safety, is certainly possible within the limits of existing techniques. Such a system need not just cater for prestige travel, but could greatly improve the services offered to the general travelling public, with both economic and social advantage.

To provide an ultra-fast system of mass transport, that is, with cruising speeds in the range of 400 to 500 km/h, beyond the capabilities of wheel-on-rail, would require the introduction of completely new technologies and new infra-structure, so that vast new investment would be needed. The development of any novel system would be a heavy burden on the resources of a single country; international collaboration in research and development would seem to be a feasible approach. By sharing test facilities and exchanging experimental findings a great deal of duplication of effort could thereby be avoided. More significantly, certainly within Europe, the system envisaged would extend beyond national boundaries, and so a large amount of international standardization would be needed in any case.

1.3   Wheel-on-rail systems

The most advanced of our conventional railway systems with steel wheels on steel rails are running at maximum speeds of little more than 200 km/h and, although it may be possible to upgrade much of the main-line networks in some of the more advanced countries so as to approach 300 km/h, there would be considerable difficulties at operational speeds much above this value. The technical limitations are primarily due to mechanical stresses, difficulties in maintaining the requisite track alignment, and, probably most serious, the use of wheel adhesion for traction and braking. The problems of adhesion represent a major limiting factor since it is necessary to allow a braking distance corresponding to the worst conditions of adhesion likely to be encountered. There are also operational problems of running a frequent service of very high-speed trains over the same tracks as other much slower but equally dense traffic. However, as the technologies of wheel suspension, track alignment, and propulsion systems continue to advance, there seems little doubt that maximum speed limits of over 300 km/h will become technically possible.[1]

Considerable effort is already being expended in a number of countries in order to extend the wheel-on-rail principle beyond the present speed limitation. Perhaps the earliest example is the now famous Tokaido line in Japan, where the 515 km between Tokyo and Osaka is covered in just over 3 hours by super-express. This first section of the Shinkansen railway system was inaugurated in 1964 and has been an enormous economic and social success; it has already carried well over a thousand million passengers without a single death or injury in accidents. However, this success has not been without serious problems. The operation of these trains at speeds of around 200 km/h creates serious wear on both the wheels and the rails, and the financial burden of maintaining the essential accuracy of the wheels and track alignment is becoming increasingly heavy. This, and the serious problems of noise and vibration associated with steel-wheeled vehicles, have been significant factors influencing the Japanese to investigate possible alternatives. Very substantial development programmes have been initiated in Japan to investigate non-contact suspensions for future high-speed travel. At the present time the extent of their research and development in the various magnetic levitation technologies and linear motor propulsions is believed to exceed that in any other country.

Radical improvements in both the rolling stock and the track of conventional railways have also been made in a number of other countries. In the United Kingdom, for example, British Rail has made spectacular advances in wheel and suspension dynamics, culminating in the successful development of an entirely new design of rolling stock, the Advanced Passenger Train (APT).[2] With a maximum design speed of 250 km/h it has been successfully tested under service conditions at speeds of over 200 km/h.

Although the speed limit of the conventional train driven by wheel traction is restricted by the frictional forces that can be developed between the steel wheels and the rails, it is confidently believed that this speed can be extended well beyond the practical limits of wheel adhesion by the use of an alternative technique, namely linear motor propulsion. With this end in view a number of large-scale developments of both the linear induction motor[3] and the linear synchronous motor are proceeding. Short-stator machines with armature winding on the vehicle and long-stator machines with a powered track winding are both being considered. With linear motor propulsion the wheels would then be used only for suspension, so that a maximum speed above 300 km/h should be quite feasible—subject of course to the limitations of track curvature, alignment, and smoothness of the rails. However, since the dynamic forces between the wheels and the rail increase dramatically with speed, extremely smooth and accurately aligned rails would be needed to ensure safe running at speeds above 300 km/h, with the consequent penalty of very high maintenance costs. Nor would the linear motor solve the noise and vibration problems of wheels which, according to the Japanese experience, become very serious above 200 km/h. Not only has it been found to be practically impossible to find effective means of solving this problem but attempts to limit the noise level by more frequent maintenance of wheels and rails are becoming prohibitively costly.

1.4 Levitated systems

As may be expected from the above discussion, present thinking[4] is that a new technology in railway development is becoming increasingly desirable, not only on the grounds of environment, economics, and passenger appeal, but also as a means of taking the critical step from speeds in the range below 250 km/h up to the 400 to 500 km/h range. Today's transport problems cannot be solved with the technical solutions of the last century. Certainly, our present-day railways and motorways cannot meet this challenge. It would appear that the key to future high-speed mass transport will lie in innovative, and possibly radical, technology. To this end a number of research groups throughout the world are investigating the technology of levitated vehicles and linear motors.

It is not expected that these high-speed 'trains' would replace existing services, but that they would add a new dimension to public transport by drastically reducing journey times between major commercial centres. For the very high-speed sections of track a completely new, straight, and level route would be essential, but in built-up areas near major stations elevated guideways could allow existing routes to be used with only minor disturbances to other services and rights-of-way. Furthermore, in comparison with new developments of road and air transport, a guided ground system would require significantly less land usage and other resources. Among the many proposals for levitated or contactless suspensions, the most promising are those using either air cushions or magnetic levitation.

The air-cushion vehicle (ACV), more generally referred to as the hovercraft, was originally developed for either land or water transport. Such vehicles can be almost independent of the surface over which they are travelling, and the most successful developments have been (i) the marine hovercraft and (ii) military craft designed for transport over mixed terrain or for use as landing craft. The initial promise of the hovercraft principle encouraged a number of groups to examine the possibility of a tracked air-cushion vehicle for high-speed ground transport.[5] In France the research has been pioneered by M. Jean Bertin since the early 1960s. In 1969 tests began on the Aerotrain, a 20 t air-cushion vehicle driven by the thrust from a ducted air propeller mounted at the rear. The vehicle was designed to carry 80 passengers at a cruising speed of 280 km/h on a straight test track, 18 km long, constructed near Orléans. Speeds between 250 and 300 km/h were attained using two 1300 h.p. turbo-engines driving a variable-pitch propeller. Further tests have subsequently been conducted on a full-scale 40-seat vehicle at Gometz which was propelled by a 400 kW linear induction motor at speeds up to 180 km/h.

Throughout the 1960s the development of the air-cushion technology was in progress in other countries also. Large-scale test vehicles were designed and constructed in both the UK and the USA. Tests were undertaken on a research vehicle at the high-speed test centre built by the US Department of Transportation at Pueblo, Colorado, and several Tracked Air-Cushion Vehicles (TACV), propelled by linear induction motors, were designed and built both for inter-city operation up to 250 km/h and for urban transport at lower speeds. In the UK a 21 t unmanned research vehicle, running on 1.6 km of elevated guideway, was built by Tracked Hovercraft Limited, and a speed of 172 km/h was reached during test runs. The special box-section guideway design used in these tests was believed to have economic advantages over the inverted-T section adopted by the French and the U-section trough of the American design.

More recently, nearly all of the development on TACVs has been either drastically curtailed or stopped altogether, mainly because of the growth of interest in magnetic levitation. Only for urban transport at relatively low speeds is there still some interest in air-cushion suspension.

Of the various types of magnetic levitation the two that are receiving most attention for high-speed operation, up to 500 km/h, are those based on electromagnetic attraction, with a small air gap or clearance between the vehicle and the track, and those involving electrodynamic repulsion across a large air gap. After a brief introduction and description of the electromagnetic and other systems, the remainder of this monograph is concerned with the technology of the electrodynamic system of levitation utilizing cryogenic magnets and linear synchronous motor propulsion. The work draws upon the results of extensive research and development programmes in a number of countries, notably West Germany, Japan, Canada, the USA, and the UK.

2

INTRODUCTION TO MAGNETIC LEVITATION

2.1 Types of magnetic levitation

The possibility of levitating ferrous objects by means of the invisible forces produced by naturally magnetized materials, the so-called permanent magnets, has stirred the imagination of inventors and inspired designers for well over a century.[6] Recently, with the development of materials technology in the post-war years, there has been a growing interest in the application of magnetic levitation to transport vehicles. Whatever the details, the basic concept is that of supporting and guiding a vehicle above its track by means of magnetic forces. As with the air cushion, the avoidance of mechanical contact allows many of the problems of wheeled systems for high-speed operation to be overcome. With the complete absence of mechanical friction, noise and wear problems may be virtually eliminated. Hence the speed of the vehicle is limited, in principle at least, only by the propulsion power available for overcoming the aerodynamic and other drag forces. Although numerous ingenious inventions have been devised to harness magnetic forces to provide stable vertical suspension, it is only in recent years, with advances in the technologies of magnetic and superconducting materials and electromagnets, that any real promise of success has been forthcoming.

Essentially, four basic methods of achieving the magnetic forces required for the suspension or support of a large vehicle have been seriously considered, namely:

   (i) permanent magnets arranged to produce a force of repulsion;
   (ii) electromagnets producing a controlled attractive force;
   (iii) a linear induction motor using a.c. power and providing both thrust and levitation; and
   (iv) electrodynamic repulsion using d.c. superconducting magnets.

More recently a fifth, hybrid, alternative has been proposed, the so-called 'mixed-μ' system which, in practice, will also use superconducting magnets.

Of these the second and fourth have so far been shown to be the most serious contenders for high-speed transport. To keep within the objectives of this series of monographs, only the latter will be discussed in detail. However, for the sake of completeness and in order to highlight the advantages, and possible disadvantages, of cryogenic levitation, all the alternatives are reviewed below.

2.2 Permanent magnets

The force of repulsion, generated continuously without the expenditure of any power, between the 'like' poles of permanent magnets is indeed appealing to the design engineer. The associated magnetic fields, used in this way, can be regarded as an invisible but permanent spring, virtually free from internal friction. Experience tells us, however, that, although a fixed magnet will repel, and can support, a magnet of like polarity above it, the supported magnet will not be in stable equilibrium. Any slight unbalance away from the equilibrium position of the upper magnet produces a destabilizing force that increases as the magnet begins to slide. In fact, on the basis that the force between two magnetic poles varies inversely as the square of the distance separating them, it was shown by Earnshaw[7] in 1842 in a mathematical theorem that a magnetic body cannot be supported in a stable manner in the field produced by any combination of magnetic poles. It follows that any stable suspension system using only permanent magnetic forces requires the movement of the suspended body to be mechanically restrained in at least one degree of freedom. For a vehicle suspended by repulsive forces between magnets on the vehicle and magnets on the track, the vertical equilibrium is stable, but any lateral movement would have to be prevented mechanically, for example by horizontally mounted wheels and vertical guides on the track. Note that as free movement along the track is needed this direction is one of neutral equilibrium.

With the discovery in the 1960s of improved permanent magnet materials, based on iron oxides, with very high coercivities and capable of being manufactured fairly cheaply in quantity, there was a renewed interest in the possibility of magnetic suspension for transport applications. Not only was the repulsion force exerted between pairs of magnets much greater than the weight of an individual magnet, but the coercivity was sufficiently high to preclude any mutual demagnetization. Investigations of permanent magnet suspensions, using barium ferrite ceramic materials, began seriously in America, the UK, and Germany, mainly with urban transport and other low-speed applications in mind. However, the weight of the magnet load on the vehicle, the cost of laying magnets along the track, and the safety requirements of maintaining the active track clear of ferromagnetic debris, have all combined to limit the interest in these materials for large-scale transport applications. Another major disadvantage is the need to provide some method of lateral guidance which, as Earnshaw's theorem shows, cannot be done by any strategic arrangement of permanent magnets. Hence an arrangement of either horizontally mounted wheels or controlled electromagnets is necessary. This would lead to a complex hybrid system with little or no advantage over the other alternatives to be described below. In the design developed by Polgreen[8] in the UK the solution using guide wheels was proposed. It should be understood, however, that the wheels would be subjected to forces comparable with those of the primary suspension, and hence the design problems would be essentially similar to those of conventional wheel-on-rail suspensions. A diagram of the scale model constructed by Polgreen, illustrating the suspension and linear motor propulsion, is shown in Fig. 2.1.

More recently a new series of permanent magnetic materials, the rare-earth/cobalt alloys, has been discovered, which have values of magnetization and coercivity far in excess of those of ferrite materials. Although they are at present only being produced in small quantities, and hence their cost is necessarily high, there is little doubt that they will find increasing application in very small-scale levitation systems.

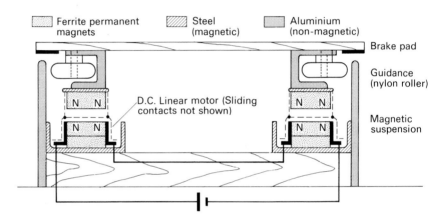

Fig. 2.1   Polgreen's permanent magnet suspension

2.3   Electromagnetic attraction

Although a ferromagnetic body cannot be suspended in stable equilibrium in a static magnetic field, stability can indeed be achieved by using adequate methods of control. By regulating the field of an electromagnet to compensate for any movement of a suspended iron body, the latter can be maintained in a stable position. Systems based on such control of the forces of attraction between vehicle-borne electromagnets and iron rails fixed along the guideway are being developed in several countries. Both stable levitation and lateral guidance forces can be generated with electronic control techniques and with suitable magnet and rail geometries.[9]

The basic principles of the method can be illustrated by reference to the graph of the attractive force versus distance characteristic of a ferromagnetic body in the field of a magnet of fixed strength (Fig. 2.2). A slight displacement from a position such as A, where the weight of the body is exactly balanced by

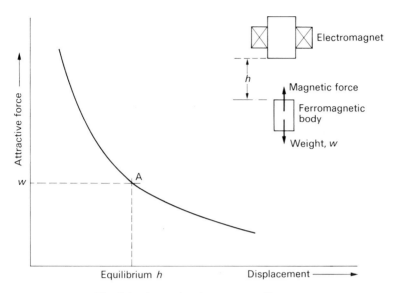

Fig. 2.2   Attraction force versus distance

the attractive force of the magnet, leads to instability. A shift to decrease the air gap produces an increase in the force of attraction; the net force is upwards, causing the body to attach itself to the magnet. Conversely, a slight increase in the air-gap distance causes the body to fall away under the force of gravity. To retain the suspended body in a stable position the field current of the electromagnet must be regulated accurately and quickly by sensing its position and/or velocity in order to maintain tight control of the air-gap separation. This idea for stabilizing a suspended body is not new, and a method for the servo-control of the magnet current was proposed at the beginning of this century. Now, however, with modern solid-state technology, a lightweight on-board controller makes the system of levitation much more appealing than hitherto. Practical systems of this type are being developed rapidly, designed around stable clearances of the order of 15 mm between the vehicle-borne magnets and the steel rails. The current in the magnet is usually controlled by a signal from a displacement transducer sensing the air-gap distance.

An alternative method of regulating the current involves feeding the electromagnets with alternating current, resonating them with a capacitor, and relying on the detuning caused by a change in the air gap to vary the current. However, detailed analyses have indicated that such a.c. systems require heavier and more costly magnets and rails than would be required by the alternative d.c. systems. A similar conclusion also holds for systems in which a.c. supplied electromagnets are controlled electronically. Hence, with advances in electronics, most development groups are concentrating their efforts on the d.c. systems.

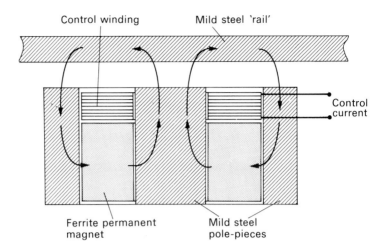

Fig. 2.3   Controlled permanent magnet levitation

A further modification of the d.c. suspension to be considered is the combination of a permanent magnet with controlled coils,[10] whereby the attractive force is provided mainly by the permanent magnets while stabilization of the air gap is attained by means of auxiliary control coils wound on the magnet pole-pieces (Fig. 2.3). In this way the net force of attraction of the magnet system is stabilized by partially cancelling or reinforcing the field of the permanent magnets. One of the main advantages claimed for a controlled permanent magnet suspension is that its operation would be fail-safe over a small air-gap range (a few millimetres), in that any failure in the power supply to the control system would still leave the permanent magnets to pull the vehicle onto the track. It could then be released by the full opposing field of the control coil when the power was restored. Subsequent design studies have shown, however, that in other respects this arrangement involves complexities similar to those of using controlled electromagnets, and that on balance it offers little or no advantage.

The magnetic attraction system has been examined in a number of countries specifically for low-speed automated urban transport systems, and a number of prototype vehicles have been constructed and tested. Pioneer work has been carried out at the University of Sussex[11] and by British Rail[12] in the UK, and passenger-carrying test vehicles have been built by both groups. Considerable development has also been undertaken by Krauss–Maffei and by Messerschmitt–Bolkow–Blohm (MBB) in West Germany and full-scale vehicles have been tested. An ambitious scheme to install a full-scale demonstration system in Toronto had to be cancelled because of technical and design difficulties found in the detail design of the suspension and the linear induction motor propulsion. There now appears to be considerable doubt about the merits of the attraction system concept when compared with conventional wheels and rails for low-speed operation.

On the other hand there has been, in Germany in particular, a renewed interest in the development of the attraction principle for high-speed inter-city transport, that is, at speeds up to 400 km/h. Although at very low speeds, less than 100 km/h, it was confidently expected that the desired ride quality could be attained by control of the primary suspension magnets, studies have indicated that this is not the case for high-speed operation. A rigid body vehicle, without secondary springing between the magnets and the load- or passenger-carrying compartment, is unable to smooth the disturbances from realistic guideways and other external agencies sufficiently well to meet the desired standard of passenger comfort. In essence the design requirement of enabling the vehicle to follow curves and changes of gradient is basically incompatible with that of isolating the vehicle from track irregularities. While the continuous power consumption needed to maintain gap clearances of from 10 to 20 mm is comparatively modest, the peak power required to adjust the lift force increases rapidly with the frequency and amplitude of the disturbances; for very high-speed operation the alignment of the tracks would therefore have to be maintained to an impracticable degree of accuracy. The alternative of very large vehicle–track clearances could allow the vehicle to absorb safely guideway irregularities with wavelengths shorter than the vehicle length and give the desired ride quality. However, such a design would necessitate unacceptably large and heavy electromagnets and their associated power supplies.

From the results of studies of the dynamic behaviour of the vehicle–guideway system, it was concluded by MBB[13] in West Germany that the ride quality requirements could be met by attaching the magnets to rigid sub-frames to form the primary suspension, the frames in turn being connected to the vehicle body by a secondary, mechanically sprung, suspension. The magnet control system is decentralized with individual controllers for the magnets, each elastically suspended magnet with its controller being a unit in a hierarchical control system. Since these units are capable of following the surface irregularities of the track they have been termed 'magnetic wheels'. If, as is believed, the magnets and their frames are able to follow the short-wavelength variations of the guideway, then lower dynamic ratings are imposed on them, and hence smaller nominal air gaps are possible. It has been predicted from these studies that gaps of between 5 and 10 mm are quite realistic for systems operating at up to 400 km/h.

In the operation of the controlled d.c. magnet suspension, particularly at the higher speeds, eddy currents will be generated in the conducting steel rails as a result of the changes in incident magnetic flux at the surface. These induced currents create a repulsion force on the vehicle magnets which tends to counteract the attractive suspension forces (it is this repulsion which is utilized in the electrodynamic suspension method described below). The induced currents, circulating in the rails, also give rise to resistive losses in the steel which are manifested as a drag force on the vehicle. To minimize these eddy-current effects at the higher speeds, and hence to reduce the propulsion power required, some degree of lamination of the track structure is necessary.

One of the major problems of the electromagnetic attraction system for high-speed running is the linear motor propulsion. Though several solutions have been proposed, a satisfactory short-stator linear induction motor design is yet to be produced. The low efficiency and power factor, together with the associated problems of maintaining a small air gap and of power collection, all tend to militate against the application of such machines as propulsion units for levitated vehicles. However, recent experiments in Germany on the long-stator version of a synchronous electromagnetic machine—that is, with a powered winding along the track—look very promising, and a large-scale, passenger-carrying, electromagnetically suspended vehicle[14] propelled by such a linear synchronous motor was demonstrated at the International Transport Exhibition in June 1979 in Hamburg.

2.4   A.C. (repulsion) systems

Repulsion systems can be divided into two categories, namely those supplied with alternating current, of which the best known is the levitating linear induction motor, sometimes referred to as the 'Magnetic River', and those using direct current magnets to produce the so-called electrodynamic levitation. In both cases the vehicle suspension forces are produced between vehicle-borne magnets and an electrically conducting strip track in which circulating eddy currents are induced. In the first case these currents are produced by an alternating electromagnetic field without the need for vehicle motion. In the second case the relative movement of the vehicle d.c. magnets over the conducting track induces the currents. This section is confined to the former case, namely the designs which have been proposed based on a.c. systems.

It is well understood that a repulsive force is created when a copper coil carrying alternating current is brought near a conducting plate. A simple explanation is that the time-varying field produced by the alternating current in the coil induces eddy currents in the surface layers of the conducting plate, which in turn generate an opposing or repulsive force on the coil. This general relationship is given by Lenz's law, according to which the currents induced in the plate flow in such a direction as to set up a magnetic field opposing the changing flux of the coil, i.e., if the magnetic flux of the coil is decaying, the induced currents tend to sustain the original flux, and, contrariwise, for an increasing magnetic flux, the induced currents flow in that direction which tends to minimize the increase in the flux. In other words, the induced currents oppose the change in flux in such a way as to avoid producing any energy additional to that which caused it. In short, Lenz's law is just the application of the law of conservation of energy.

In 1912 Emile Bachelet, a French engineer, utilized this principle in designing and building a model vehicle which was levitated and propelled by magnetic forces. In his system[15] the vehicle, carrying a conducting aluminium plate attached to its underside, was levitated above a continuous row of electromagnets, supplied with alternating current, distributed along the track. This levitation came to be dubbed the 'Foucault Railway' after the early French scientist who first investigated these so-called Foucault, or eddy, currents. It became quite obvious early on that not only would the costs for a full-scale system with iron-cored copper magnets along the track be excessively high but also there would be a huge power consumption. For example, 15 kW was required to levitate a 15 kg model to a height of 1 cm. It was then proposed that the arrangement be inverted and Bachelet designed an 0.5 t mail-carrying vehicle having electromagnets in wings attached to the body. In this design the suspension forces would be generated by the vehicle magnets interacting with a pair of continuous aluminium strips or tracks along the guideway.

Later, following advances made in linear induction motor (LIM) designs, these ideas were developed further, whereby, in addition to the horizontal forces providing thrust, the vertical forces produced by the machines were utilized for levitation. In 1971, Hochhausler[16] in Germany demonstrated a system comprising a linear a.c. field winding along the guideway interacting with a conducting plate on the vehicle to produce both propulsion and levitation. In the UK the system devised by Laithwaite and his colleagues[3] and generally referred to as the 'Magnetic River' had the stator winding of the LIM mounted on the vehicle, the levitation and propulsion forces being generated by interaction with an aluminium strip or reaction rail along the guideway. Although this suggested design led to considerable savings in track costs, as compared with the active track system, the problems of power collection and the considerable weight of the machine and its on-board power conditioning equipment could pose serious problems for high-speed transport application. More fundamental drawbacks of the Magnetic River scheme, however, result from the induction mechanism. These will be described in Chapter 5. However, with a relatively large gap between vehicle and track, as would be desirable at high speeds, unacceptably low power factors and efficiencies cannot be avoided. At high speeds also there is a fall-off in the forces generated by the motor so that it has to be operated in an inefficient mode, i.e. at high slip speed. Hence this concept does not appear to offer very promising prospects. A variation of this scheme of providing both lift and propulsion with the linear motor was attempted by Rohr Industries in the US; a demonstration vehicle has been tested at moderate speeds, but no further developments have been reported.

2.5   Electrodynamic (d.c.) repulsion

With the development of the superconducting magnet in more recent years and its commercial exploitation, a lightweight source of intense magnetic field has become generally available. Superconductivity, or the state of zero electrical resistance which occurs in certain metals and alloys when cooled to liquid helium temperatures (boiling point 4.2 K), has revolutionized magnet design by enabling large currents to be carried in relatively small conductors. For example, a copper wire of 1 mm^2 cross-section is limited to about 10 A in normal use, whereas a superconductor of the same size may be rated at up to several thousand amperes. Thus, the superconducting magnet can be designed without the heavy iron core of the conventional electromagnet and, furthermore, it consumes no power during its operation apart from the relatively small amount of refrigeration needed to maintain its low-temperature environment.

With the ready availability of such powerful lightweight magnets Bachelet's original ideas of levitating a vehicle by the forces generated by eddy currents induced in an aluminium sheet could be reconsidered. A significant difference, however, is that alternating current cannot be used for the generation of the eddy currents. The superconducting magnet is only able to tolerate d.c. currents without energy loss. Therefore, to produce the time-varying magnetic field required to induce eddy currents in the aluminium track, it is necessary to move the vehicle-borne d.c.-excited superconducting magnets relative to the track. When this is done levitation forces similar to those produced by a.c. excited magnets are generated electrodynamically.

For the purpose of discussing the historical development of this mode of levitation, using superconducting magnets, only the general principles will be outlined here, leaving a more detailed explanation of the system behaviour to the next chapter. As has been seen, the sheet conductor, in the form of an aluminium strip track, is exposed to a rapidly changing magnetic field as the vehicle travels over it and, as a result, circulating eddy currents are induced in the aluminium in such a way as to oppose the field and prevent it from penetrating through the track. In other words, the aluminium with its surface currents now behaves like a diamagnetic material excluding the magnetic flux of the vehicle magnets. As shown in Fig. 2.4, the flux lines are compressed in the air gap between the vehicle and its track. Because of this flux compression a magnetic pressure equivalent to $B^2/2\mu_0$ is created, where B is the magnetic field strength at the track and μ_0 is the permeability of free space. As an example of the magnetic pressures which are possible, for a field strength of 1 tesla, which is relatively modest by the standards of superconducting magnets, the pressure

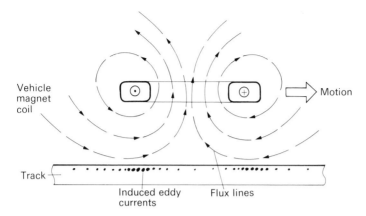

Vehicle magnet coil

Motion

Track

Induced eddy currents

Flux lines

Fig. 2.4   Electrodynamic levitation

created would be approximately 4 atmospheres and, since these pressures can be produced over a considerable area, the levitation of a 100-passenger vehicle, weighing typically 30 t, becomes entirely possible. Sometimes, for purposes of calculation, it is more convenient to explain the magnetic pressure in terms of an imaginary or 'image' magnet of opposite polarity located an equal distance below the track as the real magnet is above it (Fig. 3.1, p. 25). In common with other eddy-current phenomena there is a characteristic depth to which the currents and field can penetrate an infinitely thick track. This so-called 'skin depth' is related to the rate of change of magnetic field at a point on the track surface and to the properties of the aluminium. Typical values for a full-scale vehicle travelling at 500 km/h are around 10 mm, so the strip track need be little thicker than this.

This electrodynamic induction method of levitation represents a notable exception to Earnshaw's theorem previously mentioned, in that it does provide a stable suspension in a magnetic force field without the need for special control systems. The reason for this, as Braunbeck[17] showed in 1939 in his extension of the stability theory, is because of the introduction of an effectively diamagnetic material, i.e., the aluminium track with its eddy currents, into the field space, thereby providing the essential ingredient for stable equilibrium. The net result is a stable and resilient magnetic pressure exerted by the vehicle magnets pushing onto the currents circulating in the aluminium track.

Aluminium, like all normal metal conductors, has only a finite electrical conductivity, and hence the induced eddy currents are subject to a certain amount of resistance. The resulting ohmic or i^2r loss, where i is the current and r the resistance, manifests itself as heating in the track and also as a magnetic drag force on the vehicle which must be overcome by the propulsion motor.

Although this drag force, F_D, is quite large at very low speeds, it has the attractive feature of decreasing with increasing velocity (see Fig. 3.3, p. 27).

With good conductors like aluminium, the magnetic drag force has been calculated to be quite tolerable in a full-scale vehicle design and, in the high-speed range of travel above about 300 km/h, it is the aerodynamic drag force which becomes dominant, increasing as the square of the velocity. It is for aerodynamic reasons, therefore, that ground speeds much in excess of 500 km/h become quite uneconomic in terms of power requirements for present-day propulsion units. At speeds above this range, partially evacuated tunnels to reduce the atmospheric pressure would probably be necessary and, indeed, have been proposed.

In 1964, a feasibility study for a high-speed test facility for rocket propulsion up to 5000 m/s was initiated at the Sandia Laboratories in New Mexico. It was found[18] that conventional tracked sleds were unsuitable for tests above about 1500 m/s because of aerodynamic heating, excessive vibration and consequent damage to the track by the guide shoes. A solution was sought whereby the test vehicle could accelerate freely without physical contact with the guideway in a tunnel which was evacuated to reduce the otherwise prohibitive aerodynamic drag. An electrodynamic system of levitation based on Bachelet's original ideas of eddy-current repulsion, but using high-field superconducting magnets, appeared to offer the best solution to this problem. In the proposed design the magnets within their cryostat assemblies would be mounted on the sides of the sled-like wings, so that both levitation and guidance forces would be generated by the movement of the vehicle within U-shaped channels clad with aluminium. A diagrammatic representation of the proposed scheme is shown in Fig. 2.5. In operation the sled would be accelerated initially on a conventional track up to a speed of about Mach 2. It would then enter the evacuated tunnel through a frangible diaphragm to become magnetically suspended out of contact with its guideway and further accelerated to maximum speeds of 5000 m/s.

Although these original ideas were never actually put into practice, considerable design studies were undertaken and one of the outcomes was a proposal to apply 'maglev' for the suspension of high-speed surface transport vehicles.

Meanwhile, Powell and Danby,[19] at the Brookhaven National Laboratory in America, had proposed a number of 'maglev' schemes for inter-city transport. In one (Fig. 2.6) put forward in 1966 the rows of superconducting coils arranged horizontally along the vehicle interacted with similarly arranged aluminium coils, fixed along the guideway, so as to generate the required levitating force. The lateral guidance was provided by the arrays of vertically mounted track coils. In a later development, in 1968, rows of superconducting coils along the vehicle interacted with rows of interconnected pairs of coils along the track. The main object of this so-called 'null-flux'

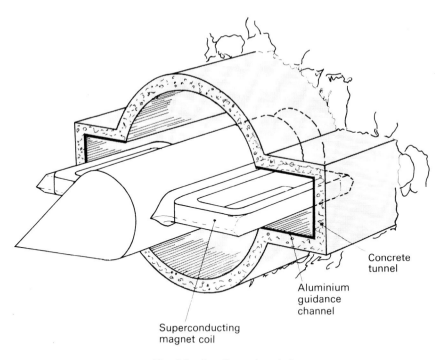

Concrete
tunnel

Aluminium
guidance
channel

Superconducting
magnet coil

Fig. 2.5   Sandia rocket sled

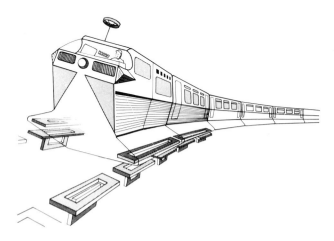

Fig. 2.6   Proposed magnetically levitated train

arrangement was to reduce the magnetic drag, as described in detail later (Section 3.2). In an attempt to reduce the inevitably high costs of such a complex track loop arrangement an alternative null-flux scheme was proposed whereby two opposing rows of superconducting coils on the vehicle would straddle a single series of track loops. By this means the track structure was simplified at the expense of the complexity of the superconducting magnets on the vehicle.

Following the work on the conducting-sheet method of levitation at the Sandia Laboratories, other groups in America undertook studies and model testing of maglev designs for vehicle transport applications. In these early designs, investigated principally at Stanford University[20] and by the Ford Motor Company,[21] the necessary lift and guidance forces are generated by the interaction between superconducting coils on the vehicle and L-shaped aluminium tracks. Subsequent to these early developments a number of variants of the original ideas began to be investigated by several research groups throughout the world. In the USA a group at MIT, sponsored by the National Science Foundation, developed to the model stage the so-called 'Magneplane', a characteristically novel scheme. Several Canadian Universities, led by a group at Queen's University, Kingston, Ontario, known as the Canadian Institute of Guided Ground Transport (CIGGT) mounted a co-ordinated programme with fairly long-term government support. In Japan a co-ordinated programme involving industry and the Japanese National Railways was set up, while in Western Germany the government supported a consortium of companies (AEG, Brown Boveri, Siemens, and Linde) in the establishment of a research group at the Siemens Laboratories in Erlangen, near Nuremberg. The main features of these various schemes are described in Chapters 4 and 6.

Since electrodynamic levitation depends on the relative motion between the vehicle magnets and the conducting track on the guideway to be able to generate the necessary lift and guidance forces, the eddy currents that would be induced at very low speeds would exclude so little of the magnetic flux that the lift force created would be less than the vehicle weight. Therefore, while the vehicle is at rest and until the necessary lift-off forces are generated (at around 50 km/h) an auxiliary wheeled suspension must be provided. This could possibly take the form of retractable rubber-tyred landing gear as in aircraft but, in any case, some means of mechanical support would be essential for any type of levitated vehicle, not solely for reasons of safety and emergencies, such as failure of power or equipment, but also to enable the vehicle to be manoeuvred in its unpowered state.

In contrast to the electromagnetic controlled levitation discussed in Section 2.3, which provides a very stiff suspension, with air gaps of around 10 mm, the electrodynamic system makes possible a stable and resilient suspension, analogous to a magnetic spring, with a track clearance between 100 mm and

300 mm, depending on the strength and size of the vehicle magnets. With such air-gap distances, small irregularities and roughness of the track surfaces would be filtered out, and likewise snow, ice, and minor debris would not present serious problems in service operation. Since the maglev vehicle would be supported over a considerable fraction of its total length, only the long-wave undulations, and hence low-frequency disturbances, would affect its ride quality from the point of view of passenger comfort. However, since magnet systems, in general, have little or no inherent damping mechanisms, some form of control or secondary suspension is required to meet the essential ride quality standards.

A further advantage of the electrodynamic system that should be emphasized is that the vehicles can be of lightweight, streamlined, monocoque construction similar to aircraft. Magnetically supported over virtually their entire length and carrying only relatively lightweight cryogenic magnets for both suspension and propulsion purposes they would not require the massive track infrastructure essential in present-day railway systems. On the contrary, relatively light and cheap guideways, perhaps elevated, and providing a minimum intrusion and obstruction to existing services, can be envisaged.

2.6   Mixed-μ suspension

In the electrodynamic system of magnetic levitation, as already seen, the eddy currents induced in the aluminium tracks by the movement of the vehicle magnets over them generate a power loss which shows up as a magnetic drag force on the vehicle. Although this force tends to decrease with increasing speed, and is dominated at high speeds by the aerodynamic drag, a significant amount of propulsive power is nevertheless required to overcome it. Another, less significant, shortcoming of the electrodynamic maglev is the necessity for some form of retractable wheel suspension for the vehicle at rest and at speeds up to about 50 km/h at which sufficient magnetic force is generated for lift-off. However, as previously mentioned, some type of 'landing gear' would always be required on grounds of safety.

Recent theoretical studies have indicated that it should be possible to design a stable suspension using superconducting magnets and iron rails.[22] This has generally been referred to as the mixed-permeability or mixed-μ system. Such a scheme would not only produce full magnetic lift at all speeds but the magnetic drag resulting from the generation of eddy currents in the iron rails could be kept small by conventional lamination or possibly by the use of sintered iron.

As mentioned above, Braunbeck's extension of Earnshaw's theorem of magnetic stability predicts that a stable levitation is possible if, included somewhere in the system, there is a material which behaves as though it had a relative magnetic permeability less than unity, i.e., is diamagnetic, and which, therefore, tends to exclude magnetic flux. A superconductor represents such a diamagnetic material since the flux is excluded because of the Meissner effect.[23] A Type I material has, in fact, a permeability (μ) virtually equal to zero (i.e., it is almost perfectly diamagnetic), and this behaviour has been spectacularly demonstrated by the classic 'floating magnet' experiment. Here a small bar of permanent magnet material is levitated in a stable manner above a saucer-shaped dish of lead cooled to its superconducting state in liquid helium. The magnetic flux of the bar magnet is repelled by circulating currents induced in the surface layers of the superconducting lead.

In the mixed-μ arrangement it has been calculated that in a system where the permeability is in some places less than that of free space (μ_0) and in others greater, that is, with mixed permeabilities, it is then possible to have static stability without the necessity for electronic control. Stable systems have been demonstrated which are made up of a superconducting material ($\mu < \mu_0$) and iron ($\mu > \mu_0$). For example, in one such experiment[22] an iron disc was statically suspended in a stable manner within the field of a superconducting coil having a superconducting screen interposed between the iron body and the coil as shown in Fig. 2.7. Although the field of the coil in the axial direction is a maximum at the centre, in the radial direction it is a minimum there. Hence an iron body could not be stably supported in this field alone. However, stability can be achieved by inserting a superconducting screen in the axial space between the iron body and the coil, as illustrated. The presence of the diamagnetic screen gives rise to a negative, i.e., stable, force in the radial direction, and since the axial force is always negative with or without the screen, the iron body is statically suspended in a stable manner in the field of the coil.

Superconducting screen   Iron   Superconducting coil

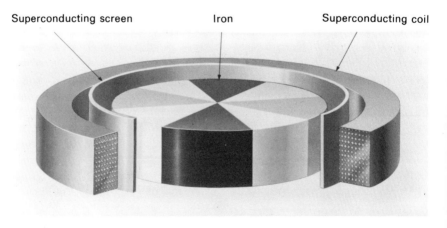

Fig. 2.7   Mixed-μ method of levitation

To be able to apply these ideas to a realistic transport system requires considerably more invention. A number of promising configurations have been proposed but more research and experiment is required before a practical design will be evolved. The most attractive features of a mixed-μ magnetic suspension are its potential for providing levitation forces at all speeds including rest, the comparatively small magnetic drag losses to be expected with laminated steel rails, instead of aluminium as in the electrodynamic system, and the elimination of the low-speed peak in the magnetic drag. A disadvantage is that the design of the superconducting magnets and screens and the cryostat assembly would be appreciably more complex than in the case of the conceptually simpler electrodynamic levitation schemes.

3

ELECTRODYNAMIC LEVITATION

3.1 Theory

Calculation of the forces generated in an electrodynamic suspension has formed the subject of many papers, of which a few are included in the Bibliography.[20, 21, 24–30] It is found that the performance of the simplest coil and guideway configurations, such as a rectangular coil over an infinite sheet, is relatively easily computed, especially in the limiting cases of low and high speeds and of thin track conductors. For realistic, but more complicated configurations containing boundaries and edges, the principles are well understood, but it proves far more difficult to obtain an explicit solution. Nevertheless, for many purposes the simple theory permits a clear description of the performance.

A first step is the calculation of the image force of a coil—that is, the force between the coil and a second similar, but imaginary, coil of the same strength, but opposite polarity, situated at the reflecting image position as shown in Fig. 3.1. This is the force generated in the high-speed limit, where the conducting sheet is assumed to behave as a perfect diamagnetic, so that the magnetic field does not penetrate the track and eddy currents are confined to a thin layer, the skin depth, at the track surface. Under these conditions the image coil system automatically satisfies the boundary conditions on the field at the track surface. In general, this force, F, may be calculated using eqn (3.1) or other standard methods:

$$F = \frac{\mu_0 I_1 I_2}{4\pi} \oint \oint \frac{ds_1 \times (r \times ds_2)}{r^3}. \tag{3.1}$$

As quoted the equation holds for any position and orientation of coil (see Fig. 3.2) and yields lateral as well as vertical forces. Evaluation is straightforward in the case of a rectangular coil, and a few sample results taken from a paper[24] by Reitz of the Ford group are given in Table 3.1.

At lower speeds the eddy currents penetrate deeper into the track conductor, they become weaker, and their effective position moves ahead of the vehicle coils. To a first approximation there is a speed above which the eddy currents are well developed, so that the lift force, F_L, is close to the image value. In this regime also the power loss due to the eddy currents is roughly constant (when the track, of thickness T, is fully penetrated by the eddy currents) or at worst varies as the square root of the speed, v, (when the eddy currents are confined to a surface layer the skin depth in thickness). Hence, by noting that the power

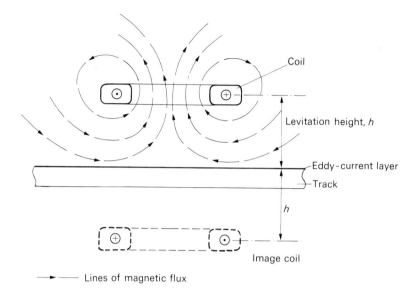

Coil

Levitation height, h

Eddy-current layer

Track

h

Image coil

→—— Lines of magnetic flux

Fig. 3.1   Cross-section showing position and polarity of image coil

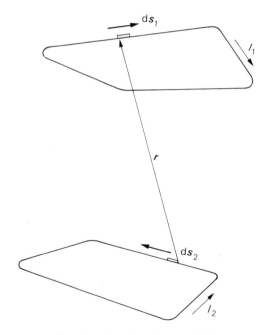

ds_1

I_1

r

ds_2

I_2

Fig. 3.2   Notation for eqn. (3.1)

Table 3.1   *Image force on rectangular coils* (a × b) *at height* h

b/a 2h/a	0.5	1	2	10
0	1.000	1.000	1.000	1.000
0.1	0.860	0.901	0.932	0.974
0.2	0.714	0.804	0.860	0.943
0.3	0.577	0.711	0.787	0.882
0.4	0.459	0.623	0.714	0.822
0.5	0.363	0.542	0.644	0.758

Multiply tabulated values by $(2 \times 10^{-7} I_1 I_2 (a+b)/h)$ to obtain the force in newtons when currents are in amperes.

is equal to the product of the magnetic drag force, F_D, and the speed, F_D may be seen to vary inversely as v or as $v^{\frac{1}{2}}$. Below the critical speed the magnitude of the induced currents will vary linearly with the speed and, hence, the drag force will also be proportional to speed; a maximum is reached, typically at a speed of around 20 km/h, at the transition to high-speed behaviour.

The variation of lift force is not so easily discussed, because the change in effective position of the image coil results in the repulsion force being no longer vertical (hence, of course, the increasingly strong drag force), but it can be shown to be proportional to the square of the vehicle speed initially. Above the critical speed current-saturation occurs so that over 90 per cent of the maximum lift is developed above 120 km/h.

The critical speed is not that at which the eddy currents just penetrate the track fully—which in typical cases is as high or higher than the service speed— but is, rather, a speed related to the rate at which the induced eddy currents decay, and is usually a few tens of kilometres per hour. These relationships are illustrated graphically in Fig. 3.3 which shows the results calculated for a 2 m × 0.8 m coil moving in the direction of its length at a height of 0.2 m over a 15 mm thick infinitely wide aluminium sheet.

Exact calculation of the lift and drag forces is possible by using Fourier methods to solve the field problem, and these have been described in detail by both the Ford group[21,25] and others.[20,27] The magnet array on the vehicle usually has coils of alternating polarity, giving rise to a wavelength, λ, of twice the coil pitch, and thus the field can be expressed as a harmonic series with fundamental period $1/\lambda$. As far as the track conductor is concerned this is equivalent to a flux density alternating in time with a frequency v/λ, and leads directly to an expression for skin depth, δ, thus:

$$\delta = \sqrt{(\lambda/\pi\sigma\mu v)}. \tag{3.2}$$

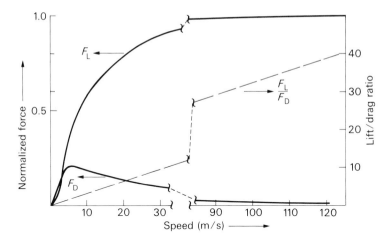

Fig. 3.3   Dependence of forces on speed

Here σ is the electrical conductivity and μ, usually equal to μ_0, is the magnetic permeability of the track.

For a single magnet rather than an array the formulation is in terms of Fourier integrals over a spectrum of wavelengths, and a simple expression for skin depth is not readily extracted. In either case, though, and for infinite planar geometries, matching of the Fourier components at the various boundaries is relatively straightforward, and explicit solutions have been obtained by the Ford group and others for the whole speed range, for all track thicknesses, and for combinations of conducting and ferromagnetic track materials.

However, for a first understanding of a practical levitation system some of the approximate results are perhaps the most useful. In particular it should be noted that there is little point in making the track much thicker than the skin depth corresponding to the service speed, so that for practical purposes only the approximate solution appropriate to thin tracks is of interest. This thin-track approximation may be obtained from the exact solution, or directly by the 'image wake' or other approximate methods of Reitz[24] and others. Under these conditions the constant of proportionality in the lift/drag ratio is best expressed as an equivalent speed w, so that the lift/drag ratio is given by:

$$\frac{F_L}{F_D} = \frac{v}{w} \tag{3.3}$$

where $w = 2/\mu\sigma T$ for $T < \delta$, T is the track thickness, and δ is the skin depth defined above.

It is also found that w is essentially the critical speed introduced above, and so the results of that discussion may be expressed algebraically by the following approximate equations derived by Richards and Tinkham:[29]

$$\frac{F_L}{F_I} = \left(\frac{v}{w}\right)^2 \Big/ \{1 + (v/w)^2\} \tag{3.4}$$

$$\frac{F_D}{F_I} = \left(\frac{v}{w}\right) \Big/ \{1 + (v/w)^2\},$$

where F_I is the image force in the high-speed limit.

It is interesting to note that, in the example of Fig. 3.3, w is equal to 3 m/s, and the skin depth to about 17 mm at a speed of 100 m/s. The remaining discrepancies between Fig. 3.3 and the approximate eqns (3.4) are probably due to the relatively small width of the coil. In this connection, it should be noted that the experimental evidence tends to favour the exact results of calculation plotted in the graph, since the Ford group, for example, were able to describe[21] their observed lift forces by a relationship

$$F_L/F_I = 1 - (1 + v^2/w^2)^{-n},$$

where n was in the range 1/5 to 1/3.

A consequence of the conditions for which eqn (3.3) was derived is that the ratio of lift to drag forces is independent of coil shape, and it thus appears that in this case of an infinite sheet track the scope for minimizing drag by suitable choice of coil dimensions, levitation height, etc., is extremely limited. Indeed, the only apparent dependence on coil dimensions is that of the skin depth, which is smaller for short coils; this decrease, however, can only worsen the drag since it reduces the effective track thickness. Another effect, not obvious from the preceding simplified expressions, is that the lift increases more rapidly with speed for short coils than for long coils; however, the actual speed at which the drag 'peaks' is not very dependent on coil dimensions and thus shorter coils exhibit a higher peak drag force (for constant levitation height).

If the track is assumed to be infinitely thick the effective thickness is roughly equal to the skin depth, and then it may be concluded, as was done above, that shorter coils show much lower lift/drag ratios (higher drag) than do longer coils. Though such a comparison has been used to justify the use of long narrow coils for levitation, it is not very relevant in practice. A more realistic procedure for selecting coil dimensions would be first to specify the maximum acceptable drag at the service speed, then to determine the necessary track thickness from eqn (3.3) and, finally, to choose the minimum coil length so as to ensure full eddy current penetration of the track at the service speed by making the skin depth (eqn (3.2)) comparable with the track thickness T. The behaviour can then be calculated at all speeds of interest on the assumption of a thin track.

The problem of the 'drag peak' remains, and must be dealt with by other measures. However, design studies have shown that, in practice, it will be confined to the very low speed range, below about 40 km/h, and several practical methods of overcoming the difficulty have been suggested. For example, in regions where the vehicle is known to travel slowly, such as when accelerating away from a station, the track aluminium may be thickened or even be omitted entirely. As higher speeds are reached, the magnetic drag force decreases rapidly so that, at the proposed service speeds, aerodynamic drag dominates.

Some of the points raised in this section are illustrated by the sample of calculated results given in Table 3.2.

Table 3.2   Computed values of F_L/F_D

Speed (m/s)	Coil length (m)				
	3	2	1.5	1	0.5
135	42	40	40	38	33
45	14.7	14.7	14.6	14.4	14.0

Coil 0.5 m wide, levitated 0.3 m above a 15 mm thick aluminium track.

In the foregoing discussion it has been assumed that all the force normal to the track makes up the lift force, F_L. In practice this will often not be the case, obvious instances being where the track is angled or has vertical walls for guidance. Eqns (3.3) and (3.4) should then be modified so that F_L is replaced by the total magnitude of the force normal to the track. If, for example, the channel sides contribute inward forces F_{G1} and F_{G2}, respectively, their sum plus the vertical lift force will appear in the equations, though the net transverse or guidance force will be $F_{G1} - F_{G2}$. As a result of this the effective lift/drag ratio can be severely degraded. There is some experimental evidence (see Section 3.3) to suggest that eqn (3.3) for $(F_G + F_L)/F_D$ is valid, at least approximately, even where the transverse force, F_G, arises from edge effects at the boundary of a semi-infinite sheet. A theoretical justification[31] has been given by the Ford group, though only for the high-speed limit, i.e. with fully developed skin effect, which will rarely be met in practice.

3.2   Null-flux systems

From the above analysis, it appears that electrodynamic levitation incurs a drag force which, for practical track thicknesses, will be at least 1.5 per cent of the vehicle weight at speeds of 500 km/h, and more at lower speeds. It must be emphasized that, at the proposed cruising speeds of several hundred kilometres per hour, this magnetic drag would be appreciably lower than the aerodynamic drag of even a carefully designed vehicle.

Nevertheless, arrangements of the track and the vehicle magnets are possible which result in a substantially reduced magnetic drag force. These are generally known as 'null-flux' systems. The original designs of Powell and Danby[32] were based on a track made up of horizontally mounted conducting loops both above and below the vehicle magnets (see Fig. 3.4). Each pair of loops was interconnected in series opposition so that the net flux linked was almost zero (hence the name); thus the current induced in the track was small, and hence the drag force was also small. Of course, to produce the required lift with such small track currents necessitates a stronger magnetic field or a greater area of active track, and the resulting requirement for more magnetic or superconducting material is a characteristic of null-flux systems.

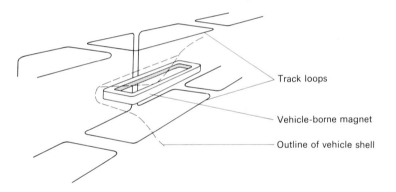

Track loops

Vehicle-borne magnet

Outline of vehicle shell

Fig. 3.4   Schematic of null-flux levitation

However, there are alternatives to a loop track. In the case of a thin-sheet track the eddy currents flow in the plane of the track, but are concentrated beneath the primary coils on the vehicles. They are related, through Maxwell's equations, to the normal component, B_n, of the magnetic flux density. Thus the drag force, which is proportional to the power dissipated by the eddy currents, must depend on an integral of B_n^2 over the track surface. The lift force is related to the product of the eddy-current density and the tangential field component, B_t, and so will depend on the product $B_n B_t$. It will be clear that if the track surface is chosen so that B_n becomes zero everywhere—in other words the track surface is defined by a flux line—the ratio of lift to drag force will become infinitely large. At the same time, both lift and drag forces will individually become vanishingly small, so in practice the vehicle will be displaced to bring the null-flux surface slightly above or below the track. The lift and drag forces then vary linearly and quadratically, respectively, with the displacement. The actual position adopted by the vehicle is determined by the magnet strength, a high strength requiring little track current to generate the required lift force, and hence corresponding to small displacement, and conversely. There is, however, a limit to the lift/drag ratio which can be attained, set by the finite thickness of the track, as has been discussed by Richards and Tinkham[29] and by the Siemens group.[27] Nevertheless, substantial improvements over the simple levitation scheme are possible. These points are illustrated further by the results of experimental work discussed in Section 3.3, p. 32.

A further consequence of any attempt to improve efficiency by reducing the magnetic drag is that the stiffness of the suspension (the proportionate rate of change of force with displacement) is necessarily altered. Stiffness is important, since it determines the natural frequency of oscillation of the vehicle, and hence it must be controlled to give the best ride performance and passenger comfort. For the null-flux systems the induced track current, I, is proportional to the field strength, B, of the vehicle magnets and is, to a first approximation, proportional to the displacement, z, of the track from the null surface. The magnetic force, F, is proportional to BI and thus to $B^2 z_0$ at the equilibrium position, and hence the stiffness, $(1/F)$ $(\partial F/\partial z)$, varies as $1/z_0$. Since the eddy-current loss depends on I^2 the ratio of lift to drag forces is also proportional to $1/z_0$—in other words, the stiffness is proportional to the lift to drag ratio. Thornton at MIT has arrived at the same conclusion[33] in a rather more general way, showing that the decay time constant for eddy currents in the track also enters into the relationship. A potentially important way around this difficulty is offered by the so-called 'hybrid' systems which incorporate ferromagnetic material.[34] The ferromagnetic attraction forces are inherently unstable and serve to make the combination more resilient without increasing the magnetic drag.

It will be appreciated that suitable 'null' surfaces must extend uniformly in the longitudinal direction, and that their number is very restricted. Examples are the vertical plane of symmetry of a simple array of horizontally arranged coils, or the plane of symmetry between two similar, but opposed, arrays. The latter was the basis of an alternative proposed by Powell and Danby, in which the track consisted of a single set of loops linking pairs of symmetrically disposed horizontal coils on the vehicle.

If, however, the magnets are so long relative to their width that their field patterns can be treated as two-dimensional (that is, if end effects may be neglected), any flux lines define a suitable surface. So far attempts to exploit this idea by shaping the cross-section of the track metal have proved rather inconclusive, possibly because it is difficult to devise a practical configuration in which the end effects will be truly negligible.

3.3 Experimental verification

There has been a considerable amount of experimental work on electrodynamic levitation for vehicle suspension, though the results are perhaps less conclusive than might have been expected. The reason for this may be that the forces generated are not very large unless superconducting coils are used in the experiments, but then there are obvious problems in making readily wide variations in the geometry.

The results of two extensive verifications of the basic theory have been published by the Ford[21,35] and Siemens groups.[36] Other groups in the USA, Japan, and more recently Canada[37] have also reported the results of experiments using electromagnets or with large superconducting magnets over rotating aluminium discs or drums, and there have been trials with test vehicles employing superconducting magnets, as described in Chapter 6 (for example by the Japanese,[38] by the Stanford University group,[39] and by the Siemens group[40]), from which some of the force characteristics of the levitation system may be deduced. In addition, other investigations have been reported which, though less wide-ranging, illuminate particular aspects of the subject.

The Ford group reported experiments with a number of superconducting coils ranging in length up to 100 mm, held above an aluminium alloy cylinder of about 600 mm diameter rotating at maximum peripheral speeds of about 140 m/s. The object was to test the theory for the normal electrodynamic suspension above an infinite sheet track, and later to investigate edge effects and guidance forces. Because the investigation of guidance forces involved cutting a channel into the drum, the radius of curvature of the coils was chosen to match that of the drum only in this latter case; consequently, in the initial experiments to test the theory, there was some uncertainty in the levitation height. It should also be noted that the rotating cylinder was solid metal, so that the 'skin depth' for the eddy currents was much less than the drum radius, except at very low speeds, and the appropriate theory was that for an infinitely thick track.

The agreement between theory and experiment was good, and the theoretical characteristics described above for the variation of lift and drag forces with speed were generally observed. The measured values of the forces agreed with those predicted from exact calculations to within 20 per cent and, in view of the many experimental uncertainties (even though these individually were thought to be small), this discrepancy may be acceptable. The Siemens group also obtained similar results, though with different sizes of coil and with a 'disc' track. The disagreement with theory was rather less, and was also thought to be explained by experimental uncertainties.

The Ford work also included investigations into the transverse forces produced firstly when a coil was brought near the edge of the aluminium drum, and secondly when the coil was moved off-centre in a channel cut in the drum. From the former experiments, modelling a coil moving near the edge of a flat-sheet guideway, it was concluded that the ratio of lift plus lateral force to drag force was independent of the lateral movement of the coil. It should be noted, incidentally, that the transverse force is directed away from the track—i.e., it is unstable. It was also found that the magnitude of the transverse force increased more rapidly with speed than did the lift force.

The experiments with a channel track were performed after machining down the solid drum used in the earlier tests. The coil width was about 30 per cent of the channel width, and the net transverse force, F_G, was the difference between the forces generated at each vertical wall. The results reported are somewhat confusing in that the ratio $(F_L + F_G)/F_D$ was found to be constant, rather than the ratio $(F_L + F_{G1} + F_{G2})/F_D$ as might have been expected from theory, but at a value marginally less than F_L/F_D for a flat sheet. Moreover, the measured lift force was modified slightly as a consequence of the limited height of the side walls. The speed dependences of the vertical and the transverse forces appeared to be similar, in contrast to the case of the flat track.

The inevitable conclusion is that significantly more experimental and theoretical work is required before a complete understanding of transverse forces can be developed.

The Siemens group work referred to above was part of a wider investigation which included experiments with 'null-flux' and other doubly-excited geometries. For these experiments the 'track' was an aluminium disc spinning with its axis horizontal, and having two superconducting coils mounted against opposite faces of the disc. Again the experimental results were in reasonable agreement (within 10 per cent) of the theoretical predictions based on an exact analysis. These results were also of interest in emphasizing the qualitative differences between the normal and the null-flux systems, which are not apparent without studying the theory in some depth. Apart from the lower drag and the greater suspension stiffness of the null-flux arrangements, the variation of the forces with speed can also be quite different. At a constant levitation height and with a thin track the qualitative difference is not great, but as the track thickness is increased extra losses arise so that the drag force can even increase at high speeds in contrast to the normal thin-sheet behaviour [compare Figs. 3.5(a) and 3.5(b)].

In the more realistic case of the lift force remaining constant (equal to the vehicle weight) and hence of levitation height varying with speed, the differences between the normal and the null-flux configurations are equally marked. Perhaps the most important difference is the stiffness which, for the null-flux system, varies by almost an order of magnitude over the speed range investigated. This is illustrated by Fig. 3.6, taken from the results of the Siemens experiments.

A number of other experimental studies have been undertaken which, while

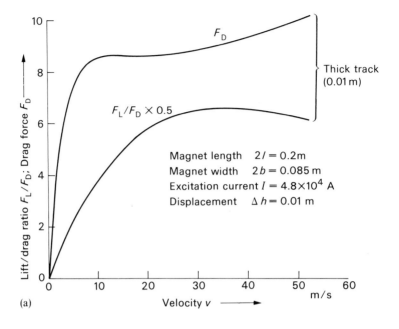

(a)

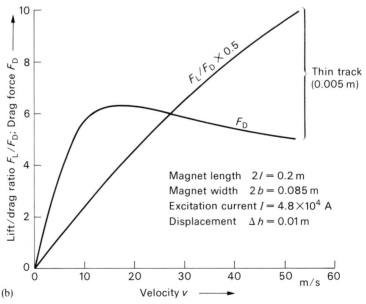

(b)

Fig. 3.5   Experimental variation of drag force with speed for null-flux suspension
(a) Thick track   (b) Thin track

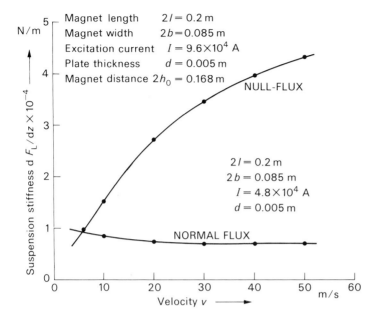

Fig. 3.6   Stiffness of normal and null-flux suspensions against speed

not precise enough to permit a quantitative comparison with theory, have nevertheless contributed to the general understanding of the levitation process. Such methods become especially valuable when the geometry of the system is so complex that analytical methods fail.

It is, for example, relatively easy to use small arrays of permanent magnets (e.g., of the rare-earth/cobalt alloys) levitated over a rotating drum or disc. Though the forces generated tend to be too small to be measured very accurately[41] the geometry can be varied more easily and cheaply than with superconducting magnets, and a qualitative impression of the relative strengths of the vertical and horizontal forces is readily obtained from such experiments.

Experiments have also been carried out using model copper coils, usually cooled with liquid nitrogen, supplied with alternating current, which can be levitated electrodynamically over an aluminium track. These readily yield a qualitative impression of the forces generated, as in the case of the permanent magnet models, but with the additional advantage that the track can be even more easily modified. Another approach, to be discussed in Section 4.2, involves making precise measurements of the impedance of the coil system in order to determine the forces. This so-called impedance modelling [18,42] has been shown to be a valuable aid to practical design.

Further studies which have been made on a.c. excited models include (i) observing the temperatures of the track surface by infra-red or liquid crystal techniques[43] in order to deduce eddy-current densities and (ii) virtually direct measurement of eddy-current densities by means of voltage probe measurements on the track surface.[37] More recently measurements of the latter type have also been made by a research group at the Eindhoven Technical University[44] in Holland, using a rig consisting of permanent magnets carried on a rotating arm and a sheet aluminium track round the circumference. The measured current densities agreed with those calculated to within 15 per cent.

3.4 Scaling problems with model systems

A final point of importance in the behaviour of models of levitation and guidance systems is that of scaling. There are two aspects which give rise to problems in practice—the one fundamental, the other practical.

The fundamental problem arises because the behaviour of the forces is governed by the characteristic speed, w (eqn 3.3), which is inversely proportional to track thickness. Thus any scaling $down$ in linear dimensions necessitates a corresponding scaling up in speed in order to remain at the same point on the force–speed curves (Fig. 3.3). Needless to say, this becomes impractical with small-scale models which, therefore, tend to be made with disproportionately thick tracks and even then exhibit low ratios of lift/drag forces. One important case where accurate scaling is possible is that in which a static model is supplied by alternating current at a frequency corresponding to v/λ (see Sections 3.1 and 4.2). Clearly the frequency now scales inversely as the square of the linear dimensions, but this is quite tolerable since frequencies are typically in the range 0–50 Hz in full-scale systems (for which wavelengths of a few metres and speeds up to 140 m/s are envisaged), and thus only in the range 0–20 kHz for a 1/20 scale model.

An additional problem is that of producing the coil systems for small-scale dynamic models. Only superconducting coils, or possibly copper coils cooled with liquid nitrogen, are able to carry the currents needed to achieve significant levitation. However, it is difficult to scale down the thickness of the thermally insulating structures for very small models. An alternative is to use permanent magnets, though it is not easy to use these to model accurately a range of configurations.

4

VEHICLE AND GUIDEWAY DESIGN

4.1   Types of guideway

As is to be expected in a subject which so readily captures the imagination, a wide variety of guideway and vehicle designs has been proposed. Nevertheless, the guideways may be discussed and summarized on the basis of the relatively small number of ways in which they have been developed from, or improved beyond, the simple flat-sheet guideway. The main objectives in design include the major ones of providing guidance as well as levitation, and of reducing the magnetic drag, together with the minor ones of permitting route switching, ensuring freedom from accumulated debris, allowing some measure of compatability with other transport services and allowing for both propulsion and levitation within the narrowest possible track.

Proposals for reducing drag in the guideway include the idea of replacing the conducting strips by a series of closed conducting loops or, alternatively, by a track in the form of a metal 'ladder'.[26, 45] These arrangements are shown in plan in the diagram (Fig. 4.1). In either case the size and pitching of the track loops may differ from those of the superconducting coils on the vehicle. In the case of the loop track, terminals are also shown, which permit some external control of the currents in the loop (for example, two tracks may be overlaid, and route switching undertaken—i.e. one selected by open-circuiting the loops of the other).

A number of analyses (e.g. Reference 45) of such track designs were made, particularly in Japan, and favourable ratios of lift to drag forces of 80 or more were found, especially if the time constant of each conducting loop was augmented by adding external inductance. Unfortunately, the basis of comparison with conducting-sheet tracks was not always made explicit, and published results for copper loops of possibly quite large cross-section have tended to be compared with those for thin aluminium tracks; a closer examination shows no clear evidence that ladder or loop tracks make much better use of the track material than conducting-sheet guideways.

A particular disadvantage of the ladder or loop track is that there is a pulsating component of the lift force which is also associated with a pulsating reaction and, hence, with a.c. losses in the superconductor. This has been analysed by a number of authors and, although by careful design the force fluctuations can be minimized (to as little as 5 per cent of the lift force) it cannot be eliminated. It has also been shown theoretically by the Canadian group that the destabilizing transverse force on the track will be worse than on a sheet

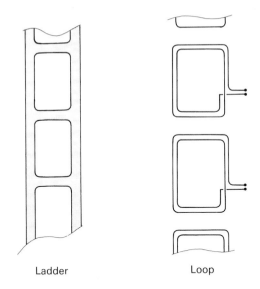

Fig. 4.1   Plan of ladder and conducting loop tracks

track containing the same amount of aluminium. This, the extra complication of building such a track, and the fact that skin effect in the conductors may degrade the performance further, are additional drawbacks.

Although, for these reasons, ladder tracks are not very appealing, it might nevertheless be expected that other methods of controlling the eddy-current paths could lead to reduced magnetic drag. One such scheme for the continuous strip track is the thickening of the strip cross-section in regions of greatest current density, as suggested by the MIT group. Although, in principle, this would be easy and cheap to implement, it does not appear, from simple impedance measurements (see Section 4.2), to offer any very significant improvements.

The classic method of drag reduction is, of course, a 'null-flux' configuration. As originally proposed the guideway was too complicated to be readily acceptable but, as was shown in Section 3.2, the null-flux behaviour can be realized in many other geometries. It remains, however, for detailed proposals to be made and, even if they are, they will almost certainly be more complicated than the schemes now current, and they will probably require considerably more superconductor for levitation than the simple sheet track. Furthermore, the increased suspension stiffness of null-flux systems will make the duties of the secondary suspension much more severe.

The problem of guidance can be solved in a number of ways. Conceptually, the simplest arrangement is vertical conducting strips on the guideway and vertically-mounted magnets on the vehicle, producing the lateral guidance

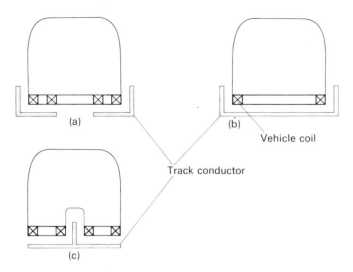

Fig. 4.2   Cross-section of vertical guide-wall configurations

forces in exactly the same way as the levitation forces are produced over the horizontal guideway conductors. Separate magnets may be used for the two functions and, although the resulting amount of cryogenic equipment may seem extravagant in a fully engineered design, there may be advantages for an experimental vehicle in having independence between levitation and guidance. It is, however, usually envisaged that the same horizontally-mounted magnet would provide the field for both functions. The final choice of guideway configuration is also influenced by the need, if any, to make provision on the track for the armature winding or reaction rail of a linear motor. Some possible track designs are sketched in cross-section in Fig. 4.2.

Other levitation schemes employ angled surfaces, though now the lift and guidance forces can no longer be independent. Typical arrangements are sketched in Fig. 4.3. There are, however, serious problems in ensuring quasi-static stability—that is, stability of the moving vehicle in roll and sway. The suspended variants, as in Fig. 4.3(c), are always quasi-statically stable, but if a linear relationship between magnetic force and distance is assumed it is found that the trough and inverted-V [Fig. 4.3(a) and (b)] are only stable if the centre of gravity is impractically low (below A in the sketch). On the other hand, such experimental evidence as there is does not entirely support this conclusion, perhaps because in reality the forces vary in a highly non-linear way with distance. However, taking all points into consideration, tracks of this type are not particularly favoured.

Rather different are guideways which use edge effects in the track structure. These include the MIT Magneplane[46, 47] [Fig. 4.4(a)], the split-track scheme

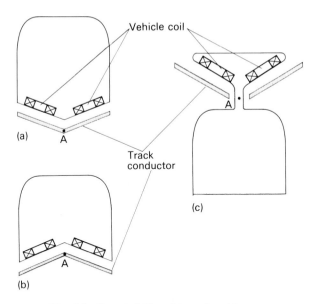

Fig. 4.3   Inverted-V and trough guideways

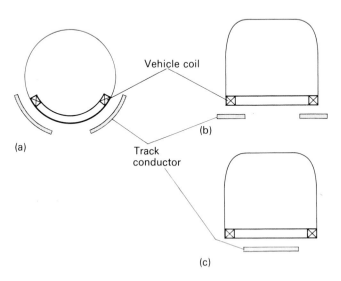

Fig. 4.4   Edge-effect guideways: (a) Magneplane (MIT), (b) University of Warwick, (c) Complement of (b)

[Fig. 4.4(b)] of the University of Warwick group,[48] and its complement [Fig. 4.4(c)], which is one of the designs studied theoretically by the Canadian group. In the first two the magnetic field penetrates through the central gap in the track, and was likened to a magnetic 'keel' by the MIT group. Nevertheless, considerable differences exist, for the Warwick design relies entirely on the 'keel' for lateral guidance as well as roll stability, whereas in the MIT design, with its wide, thin 'pancake' coils and curved track, a substantial part of the guidance must arise from the normal reaction on the levitation strips. The main merits of these designs lie in their simplicity, the economy of usage of the track aluminium, and the natural provision of space for the powered track winding of a linear synchronous or d.c. machine for propulsion [Fig. 4.4(a) and 4.4(b)]. However, although a small-scale model of the Magneplane has been successfully demonstrated,[49] although the Warwick proposals have been extensively investigated by impedance modelling[50] so that the dependence of the lift and guidance forces on track geometry is fairly well understood, and although the Canadian group have shown that there is only a limited range of (small) levitation heights over which the third configuration is stable, nevertheless much still remains unknown about these schemes, including such important characteristics as the precise speed dependence of the forces. Moreover, because of the strong interaction between lift and guidance forces, careful design of the guideway is necessary for optimum performance.

A final class of guidance system is that being developed by the Canadian[51] and German[52] groups, in which figure-of-eight loops laid across the track have a null-flux interaction with the vehicle magnets providing the field for the linear synchronous propulsion; levitation is by a quite independent set of magnets over aluminium strips along the edges of the track, as shown in Fig. 4.5. The main advantage claimed for such guidance is the low magnetic drag force associated with it. In addition, and in the same way as for the ladder tracks described above, steering or track switching is in principle made possible by open-circuiting loops to make them inoperative.

4.2   Estimation of force characteristics

There are three ways currently being used to obtain the magnetic force characteristics of a guideway, viz (i) computation, (ii) scale modelling, and (iii) impedance modelling.

Of these computing has been little used, mainly because useful results can only be obtained by working in three dimensions, and hence very heavy demands are made both on computer storage and on programming effort. Despite this some progress has been made[30] by the Canadian group who have, for example, published a number of computed diagrams of eddy-current paths

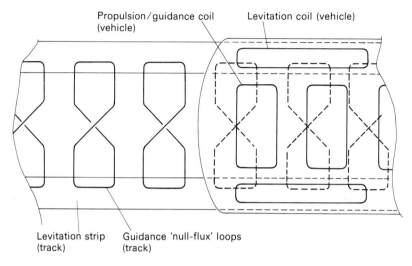

Fig. 4.5   Null-flux guidance in plan

for the geometry of Fig. 4.4(c), while other groups are known to be working on the problem.

Direct measurements of the forces generated have also been made. The most useful published results, such as those of the Ford group,[21, 35] have been obtained using a superconducting magnet supported above a rotating track, such as the circumference of an aluminium drum with its axis horizontal. However, some work has also been carried out using rotating discs (with axis vertical) or linear tracks. Such models have tended to be rather inflexible, serving as a basis for extrapolation and for comparison with theory rather than as scale models of proposed revenue systems. The MIT Magneplane model[49] was an early attempt at scaling but, as we have seen earlier (Section 3.4), true scaling is quite difficult to obtain since, as the linear dimensions are reduced, the speed has to be increased. Therefore with small models one tends to work with disproportionately thick tracks and high magnetic drag. Recently large systems and test vehicles have been operated in several countries. Because of their size such experiments are rather inflexible and represent the culmination of the design process rather than an early stage in it.

For very small models permanent magnets can be used, but the forces become so small that accurate measurement is difficult. Such models are therefore most useful for demonstrating the basic principles of levitation with relative ease and at low cost.

A widely used static technique for the estimation of magnetic forces is that of 'impedance modelling',[42] where the apparent impedance to alternating currents of a model coil positioned above a conducting-strip guideway is measured. The force on a coil is related to the current in it and to the gradient of the inductance and so, by measuring the impedance as a function of the displacement of the coil, the vertical (lift) and transverse (guidance) components of the gradient of inductance can be found; in general there will be no longitudinal component, except at joints in the track. Of course, the modelling is only useful if it can be assumed that the eddy currents induced in the model track with the static coil will behave in a manner sufficiently analogous to those generated in the real track by a moving coil; this is most nearly the case when the model supply frequency, f, is related to a wavelength, λ, equal to twice the pole pitch, and speed, v, of the real vehicle by

$$f = v/\lambda.$$

This presupposes that the spectrum of the moving magnetic wave is dominated by the wavelength λ and, although this is probably a reasonable assumption for an array of magnets, it may be only a very rough approximation in the case of single magnets and of edge effects. Moreover, in the real case, there is a longitudinal displacement between the moving magnets and the eddy-current pattern in the track which only approaches zero at very high speeds; in the a.c.-excited analogy, on the other hand, there is a corresponding phase displacement between the induced eddy currents and the exciting current, which also approaches zero as the frequency becomes very high. It is, nevertheless, found that for speeds or frequencies above the drag peak, and for the simple case of a wide-sheet guideway, the agreement between the model and the real guideway is good. Moreover, the resistive component of the impedance measured in the model can be used to estimate the magnetic drag in the real case.

The impedance modelling technique is particularly powerful because, having scaled the dimensions, it is easy to compensate by increasing the frequency, and it is found that, in such relatively cheap and easily constructed models, the measurements can be made in the very convenient audio-frequency range. As an example of this method, if the coil has inductance and resistance L_a and R_a, respectively, measured at the appropriate frequency, and a resistance R_0 when measured away from the track, then the lift, F_L, and drag, F_D, forces when it carries a direct current, I, are given by:

$$F_L = \tfrac{1}{2} I^2 \operatorname{grad} L_a$$
$$F_D = I^2 (R_a - R_0)/v$$

4.3   Guidance forces

Before discussing the stability of the vehicle on the guideway it is appropriate to consider the characteristics of the magnetic forces set up in the different guideway designs.

The lift force, discussed in Chapter 3, must obviously equal the vehicle weight, and the strength of the magnets must be adjusted to satisfy this condition at the desired levitation height. As has already been shown, the lift force varies as the square of the magnet strength, and also changes rapidly with height. Examples of the latter characteristic are given in Table 3.1 (p. 26), and a typical one is shown in Fig. 4.6. For comparison with the guidance forces, and for the analysis of subsequent sections on stability, the suspension stiffness, K (the slope of the force–height characteristic), is perhaps the most useful design parameter, expressed either directly or as a natural frequency of oscillation $[=\sqrt{(K/\text{weight})}]$.

The guidance force forms an altogether wider subject, partly because of the range of track geometries involved, and partly because of a lack of either a complete theory or comprehensive experimental evidence. The discussion following will therefore centre on specific cases for which some details have been published.

In the case of channel guideways (Fig. 4.2) as originally proposed by the Ford group, the lateral force was found to be about 10 per cent of the lift force when the clearance between the coil and the channel wall equalled the levitation height, and rose to between 20 and 40 per cent as the clearance was reduced by one-third. The corresponding lateral stiffness (about 10 per cent of

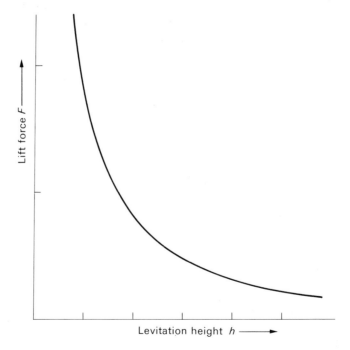

Fig. 4.6 Typical lift-force versus height for electrodynamic suspension

the vertical stiffness) could undoubtedly be increased by reducing the clearance, but the penalty for doing this would be an increase in the eddy-current losses. The Stanford Research Institute test vehicle,[39, 53] which was levitated over a channel guideway made of two L-shaped sections, as in Fig. 4.2(a), by four magnets at the corners, had a lateral stiffness of about 20 per cent of that in the vertical direction.

Other track geometries for which the guidance has been studied in some depth include designs proposed by the Canadian Maglev Group,[37, 51] the Japanese National Railways,[38] the University of Warwick,[48] and MIT,[47] respectively. The principal guidance mechanism in the Canadian system is based on closed, conducting, 'figure-of-eight' loops arranged along the track, as sketched in Fig. 4.5, which interact with the propulsion magnets in a null-flux manner. The principle of the Japanese scheme is similar; however, the guidance is combined with the armature winding of the linear synchronous motor, as described in detail in Section 6.3. With the vehicle centred over the track, the net flux linkage is zero, and hence no power loss arises from the guidance. Any lateral displacement of the vehicle induces current flow in the loops, in such a direction as to give rise to a restoring force. The lateral stiffness was designed to be somewhat less than the vertical stiffness, and the way the guidance force depended on the forward speed of the vehicle was arranged to be compatible with the speed dependence of the lift force. Both conditions were easily met with loop conductors of about 100 mm^2 cross-section. Moreover, these properties can be varied quite independently of the other properties of the track, for example, by changes in the loop size, spacing, or conductor size. In the design finally proposed, the maximum lateral force developed by the null-flux loops is about 40 per cent of the vehicle weight. However, at the displacement corresponding to this the propulsion magnets begin to interact with the levitation strips to produce an additional centring force. The total lateral force can be designed to exceed the vehicle weight, which is regarded as essential from the point of view of safety. This additional force is similar to that of the primary guidance force of the Warwick scheme [Fig. 4.4(b)], although now arranged for there to be negligible force and negligible associated losses with the vehicle in the central position. In practice, however, the vehicle will always be subject to side forces of one type or another, and it is a nice design point whether the resultant average power loss is less with the null-flux loop guidance scheme than with the split-track scheme, given that for the same total quantity of conductor on the track the levitation strips in the latter scheme can be somewhat thicker. A similar loop guidance scheme has been tested recently by the Siemens group on their test track at Erlangen.[40]

The MIT Magneplane 'magnetic keel' and the Warwick split-track designs have also been extensively investigated, both analytically and by means of impedance modelling techniques. As might be expected, the forces are strongly dependent on the track geometry and, moreover, the lateral force is associated with a rolling moment and a change in lift force. This is also, to some extent, true of null-flux guidance, though a direct comparison of the two systems is not yet possible. Typical results from impedance modelling for the Warwick track are shown in Fig. 4.7, though it must be emphasized that these are only illustrative, and they could assume quite different values as a result of small changes in geometry. Because of the complexity of the analysis of this problem, there are still many aspects not completely understood, such as the important one of speed dependence of the forces. It appears, however, that it is relatively easy to determine a geometry in which adequate lateral stiffness—that is, of the order of the vertical stiffness—together with a maximum lateral force of the order of the vehicle weight can be achieved. For the Warwick split-track design there is also some evidence that a favourably low drag force also results for at least some combinations of levitation height and track width.

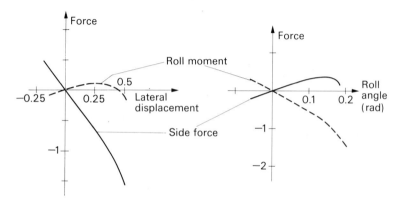

Fig. 4.7   Magnetic side-force and roll-moment versus side-slip and roll

4.4   Vehicle stability and aerodynamic forces

4.4.1   *Statement of problem*

Up till now the design and main characteristics of the magnetic lift and guidance forces on the vehicle have been considered. However, the design of a passenger vehicle requires a deeper understanding of the effect of these forces on the vehicle, viz how they vary with vehicle position and attitude and how they interact with externally imposed forces. The three main aspects to be considered are firstly, the obvious one of ensuring that any external forces to be expected should be unable to force the vehicle off the track; secondly, the need to ensure that uncontrolled oscillations of the vehicle cannot develop, or, in other words, to maintain dynamic stability; and thirdly, to make certain that the motion is sufficiently free from jolting and jarring to be acceptable to passengers.

Clearly, in discussing these matters magnetic forces are only half the story. The other half concerns the external forces which disturb the vehicle, namely those arising from aerodynamic effects, from the inevitable roughness and misalignment of the track, and from deliberate departures from a uniform motion, such as cornering, accelerating, and braking. Although the whole subject of ride quality has been extensively studied, a great deal of work remains in order to translate the results into a definite design of passenger vehicle. Part of the problem lies in going beyond the stage of estimating the vehicle response to a specific well-defined input, such as a step in the track or a wind gust, and allowing for the random nature of the natural forces. Although the analysis of these responses is not difficult, the basic data, such as the degree of jolting acceptable to passengers or even the best method of specifying this, are somewhat open to debate. What follows will therefore necessarily appear inconclusive.

4.4.2   Aerodynamic forces

In the discussion of propulsion in the next chapter, aerodynamic drag on high-speed vehicles will be considered. Here we are concerned with those aerodynamic forces which are normal to the direction of motion, and with aerodynamically generated turning moments. Most of the forces on the vehicle are generated by cross-winds or, what is equivalent, as a result of yawing or pitching of the vehicle. It is usual to normalize the forces with respect to an air 'pressure' in the direction of motion of the vehicle, $\frac{1}{2} \rho A v^2$, where ρ is the density of air, A is the cross-sectional (frontal) area of the vehicle, and v the air speed, with the result that the normalized forces are essentially independent of speed. The forces are also found to be either reasonably constant or to be linear functions of pitch and yaw for small angles, typically less than 0.2 rad yaw. Since 0.2 rad corresponds to a cross-wind of 27 m/s when the vehicle speed is 135 m/s, it will only be reached, in practice, during exceptionally stormy conditions. It should be noted, however, that careful design may be needed to keep all force components, including the aerodynamic lift, within reasonable bounds over this range of conditions.

Values for the aerodynamic forces can be obtained fairly easily from wind-tunnel tests on small-scale models. Using such data together with the corresponding data on the magnetic forces, e.g. from impedance modelling, the motion of the vehicle can be analysed and its stability investigated. Such an analysis has been carried out by the group at Warwick University,[50] who have been investigating the split-track design. A particular feature of this track geometry is the rather strong coupling between the magnetic forces in the side-slip, roll, and yaw modes. However, many of the results are of more general relevance. As the lateral motion is, to a first approximation, independent of the vertical (heave and pitch) motion the analysis can be simplified by considering the two separately. It is found that the conditions for quasi-static stability—that is, the only vehicle movement is a uniform longitudinal motion—are fairly easily satisfied. Moreover, as has been seen above, the magnetic guidance forces can be made large enough to prevent all but freak conditions from derailing the vehicle.

4.4.3   Cornering

For the greatest possible passenger comfort there are two special cases that have to be considered. One is that of cornering, where a combination of banking of the track and rolling of the vehicle should largely counteract the centrifugal force on the passengers; the other is that of cross-winds, which should ideally introduce no additional roll movement. Both conditions can be satisfied, at least for the case of zero-banking, which is the only one that has been studied though it may also be the most severe.

However, when the dynamic behaviour is analysed the range of allowable parameters which satisfy simultaneously both the stability and the foregoing 'ride comfort' criteria, becomes very small, and there is clearly scope for a great deal more design effort. Even the stability analysis under dynamic conditions can only be regarded as preliminary, based as it is on linear approximations to the the force characteristics. Another aspect of the analysis which cannot be taken fully into account is that of damping, mainly because of inadequate data. It is known, for example, that the inherent damping of the magnetic forces is very small, although it is possible to envisage some lateral damping arising from the drag force associated with the lateral movement of the vehicle over the track conductors. Estimates can also be made of aerodynamic damping. In the case of some modes, such as heave and yaw (the latter, particularly, if a tail fin is added to the vehicle), the damping can reach a significant fraction of the critical value. On the other hand, for the roll and, perhaps surprisingly, the side-slip modes, the damping seems to be quite inadequate. This has been recognized since interest in magnetic levitation first developed, with the result that the analysis of passenger comfort and the provision of additional damping is an extensive subject.

4.5   Ride quality

There are three main approaches to the provision of additional damping. One is by means of the linear synchronous motor drive, where this form of propulsion is used, one is by means of a secondary mechanical suspension, either actively controlled or passive, and the third is by means of either active or passive auxiliary magnet coils. In the first method, as studied by the MIT[47] and Canadian[54] groups, the vertical force developed by the linear motor can be varied independently of the thrust, and hence closed-loop control of the heave motion of the vehicle is possible. This control is achieved by varying both the supply current and the power angle (the phase angle between the travelling magnetic field and the vehicle field). It is by no means clear how far this can help in damping modes other than heave, though observations of the MIT model vehicle show there to be a noticeable improvement in ride quality as the control loop is closed.

Damping can otherwise be provided by a damped mechanical linkage, i.e. a secondary suspension between the main levitation magnets and the passenger compartment. Such a suspension may incorporate air-springs. Short-circuited coils or metal plates, such as the walls of the cryostat, interposed between the levitation magnets and the track, can also provide damping by dissipating the induced energy. Yet other alternatives proposed include the use of auxiliary magnet coils, or the main coils themselves, or air-springs connected into an active control loop.

Specification of the degree of damping required is usually arrived at by considering the response of the vehicle to random disturbing forces arising, for example, from track misalignment, though it could equally well result from wind gusting. Assuming the 'roughness' to be the only source, it can be expressed as a power spectral density (PSD) function, the frequency range of interest being related to misalignment over distances of a few metres to a few hundred metres in a way which, in practice, can be taken as one of a few simple forms. For example, A/Ω^2 is often used, where Ω is the wave number and A is a constant characteristic of the track, or this expression may be modified slightly to avoid the singularity at $\Omega = 0$. As far as the vehicle, moving with speed v, is concerned, the disturbance appears to have an angular frequency $v\Omega$. Applying this input to the suspension yields the output power density spectrum which is experienced by the passengers.

Many criteria have been proposed for what constitutes a comfortable ride, and no single standard has emerged as the accepted norm. The problem is obviously difficult, since many other factors can influence passenger reaction than simply the amount of jolting. These include, for example, seat design, the travel time, and whether the service lived up to expectations. Nevertheless, much of the published work refers to standards originally proposed for an air-cushion vehicle for urban service in the USA, the so-called UTACV,[55] shown in Fig. 4.8. Similar standards have also been proposed for the lateral acceleration. These set design limits on the acceleration to be experienced by the passengers as a function of frequency.

A second consideration, apart from passenger comfort, is that, whatever motion results, the vehicle must always stay clear of the track. Although obvious, this requires that the displacement of the primary suspension (the levitation magnets) be calculated also, using similar methods. Needless to say, the conditions for acceptable passenger acceleration, on the one hand, and

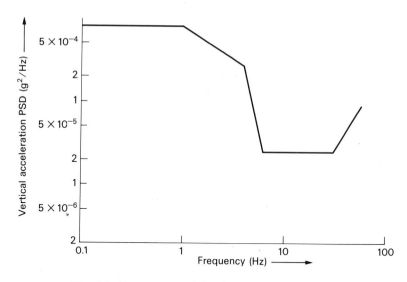

Fig. 4.8   Power spectral density proposed for UTACV

primary suspension displacement, on the other, are not necessarily compatible.

Theoretical studies of mechanical secondary suspensions have been made by several research groups, and most of these lead to the conclusion that, at least as far as vertical motion (heave) is concerned, it should be possible to meet the UTACV standards, though the range of permissible suspension parameters turns out to be rather small. An example of such a theoretical configuration, and the computed response, are shown in Figs. 4.9 and 4.10. The problems in design are most severe around the peak at about 1 Hz (the natural frequency of the magnetic suspension). Fortunately reduction of this to an acceptable level does not lead to intolerably large displacements of the magnets themselves.

A slight refinement of this analysis is to allow for the fact that the guideway disturbances are applied, not at one point, but over the length of the levitation magnets which, in many instances, is virtually the whole length of the vehicle. The result is a smoothing at all wavelengths comparable to or less than the vehicle length. This effect has been analysed[56] by the MIT group, who find a strong attenuation above a few hertz which brings the response even further within the UTACV curve. However, for some modes there can be a slight increase in amplitude of the 1 Hz peak.

The alternative of electrical methods of damping control has also been considered as part of most of the major studies of magnetic suspensions. Passive electrical damping might be expected to be relatively simple and cheap to implement, and to be inherently reliable. Unfortunately, however, it is not

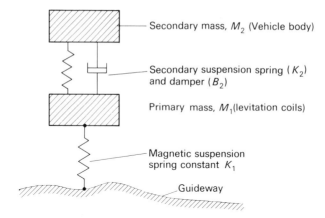

Fig. 4.9   Scheme of secondary suspension analysed

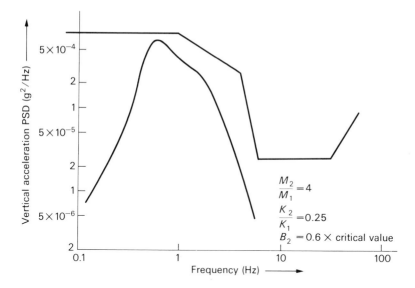

Fig. 4.10   Vertical response of secondary suspension

practicable to obtain sufficient damping at the very low natural heave frequencies of the vehicle, except perhaps in very special cases, or if the ride quality standard is sufficiently relaxed. Nevertheless, passive damping brings other benefits, in screening track-induced field and current fluctuations, especially at the higher frequencies, from the main levitation coils. Therefore, it is likely to be implemented in any full-scale vehicle. Several forms of passive screen have been proposed, such as a 1 cm thick sheet of aluminium at normal temperature (as might be provided by the cryostat casing itself), or a 1 mm sheet of copper cooled to liquid nitrogen temperatures (which might, in addition, serve as a thermal screen in the cryostat). These have been calculated to introduce damping time constants of the order of two seconds. Alternatively multiturn short-circuited coils have been used in place of conducting plates, as on the Stanford[39] research vehicle. Yet another variant, proposed by the Ford group,[21] is to resonate the damping (or screening) coils by means of capacitors, in order to enhance the energy extraction at the vehicle's natural frequency and hence to reduce the peak of the response curves.

Active damping by means of a closed control loop, on the other hand, offers much more scope for the control of the ride quality, and indeed the predicted performance of some theoretical designs is well within the UTACV curve. Such a scheme has also been implemented successfully on an experimental vehicle built by the Siemens group[52] in Germany. This performance can be achieved with modest current excursions, with, for example, peaks up to 5 per cent of the main levitation current; nevertheless these small currents imply ratings for the output stages of the control amplifiers of up to several hundred kVA and, therefore, such equipment would add significantly to the volume, the expense, and the on-board power requirements of the vehicle.

4.6 Superconducting magnets

Superconducting magnets play a vital role in electrodynamic systems of levitation, since it is only with superconductors that the necessary combination of levitation forces and large track clearances can be realized at reasonable cost. At the same time, the conceptual design of the suspension and propulsion systems can be carried out to a large extent independently of the detailed problems of the design of superconducting coils, their cryostat assemblies, and their refrigeration systems. Therefore only a brief summary of the main points of design will be given here.

As far as the vehicle magnets are concerned, the required fields (a fraction of 1 tesla at the track but reaching a few tesla at the magnet windings), total excitation (up to 10^6 ampere–turns per coil), average current densities (a few thousand amperes per square centimetre), and typical coil dimensions (such as $3 \, \text{m} \times 0.5 \, \text{m}$), are all well within the proven capabilities of modern superconducting materials. There are, however, a number of problem areas where further work is needed, though none are intractable, as has been amply demonstrated by the number of large-scale levitation magnets which have been tested to date under realistic conditions. Specific points of importance in the design are the mechanical stressing of the magnets and the possibility of heat dissipation due to time-varying magnetic fluxes.

The stress problems arise primarily because the coils that have been designed for levitation are not circular. Indeed, for the ideal design they are rectangular with straight sides, and any rounding of the corners is found to degrade their performance (this effect is only slight for wide tracks, but much greater for the split-track geometry). Thus appreciable stress concentration tends to occur at the corners, although the amount of this depends on the details of the coil construction, e.g., the degree of mechanical support given to the straight sides, the extent of the rounding of the corners, the use of resin impregnation of the coil, etc. Indeed, one aspect of interest in the proposed full-scale designs is the technique adopted for tying together the opposite sides of the coils. Though there is wide agreement on what needs to be done, much practical experience will be needed to establish the relative merits of the various competing proposals.

A.C. power losses can occur in the coils due to magnetic field fluctuations. These may arise at low frequencies from irregularities in the motion of the vehicle and/or in the track, or at power frequencies from armature reaction of the linear motor with powered track winding. In either case estimated magnitudes of the field fluctuations at the superconductor are small, especially when screening due to the metal walls of the cryostats and other structures is allowed for. Nevertheless the resultant losses could cause unacceptable temperature rises in badly cooled coils. As Hunt[57] and others have pointed out, the a.c. field amplitudes are so small that on these grounds superconducting composites with a solid core may be preferable to the multifilamentary conductors. However, the latter may still be preferred in order to ensure stable and reproducible operation of the magnets.

Although there is no reason to suppose that a technically and economically acceptable form of construction cannot be found, more design studies and operating experience are required before a satisfactory commercial design can be completely defined. To maintain the low temperatures, near the boiling point of liquid helium (4.2 K), needed for superconduction, the coils are contained in a helium cryostat, which is traditionally a multiwall vacuum-insulated vessel. A severe constraint on the design is posed by the requirement that the largest possible fraction of the levitation height of the coil should be the actual clear space between the vehicle and the track. One consequence of this is to favour badly cooled constructions, such as a resin-impregnated coil bonded to cooled backing plates, rather than coils fully immersed in liquid helium, and thus to exacerbate the coil design problems. In any case, the realization of flat-walled vacuum vessels to match the track cross-section, yet having minimum overall thickness and weight, is a difficult design problem. Several novel approaches to this problem have been published. For example, the Stanford group[39] used compressed glass-fibre mats as insulating supports between the walls of the vacuum vessel. Nevertheless, it seems unlikely that the weight distribution of levitated vehicles can differ much from that of other forms of transport where, typically, around half of the gross weight is attributable to the complete suspension and propulsion systems.

The remaining relevant aspects of cryogenic engineering is that of refrigeration, with the choice lying between an on-board refrigeration plant, on the one hand, and, on the other, a cryostat capable of maintaining the low temperatures without external services for periods of at least 8, and preferably 24 or more, hours.

Objections to the on-board refrigeration plant include (i) the additional weight, though significant progress is being made in developing lightweight compressors and refrigeration systems, (ii) the extra complexity and, therefore, maintenance requirements and cost, and (iii) the additional on-board power required, which would not be significant if the prime mover were also on board, but otherwise could easily double the auxiliary power to be supplied to the vehicle for on-board services.

The alternative to a closed refrigeration system would be a sealed cryostat. Since the loss of helium from a cryostat vented to atmosphere could not be tolerated, and since the volume of helium would render storage at atmospheric pressure impractical, storage at high pressure would be necessary. However, for gas at ambient temperature, this would require a compressor, and thus, to some extent, it would incur the disadvantages of on-board refrigeration. Fortunately, both liquid and gaseous helium have quite substantial thermal capacities when compared with most other substances at the temperatures in question, and it has been shown that, if the cryostat is sealed, the temperature and pressure will rise steadily but slowly.[58] Up to the critical point of helium, at 5.19 K and 2.26 atmospheres, the cryostat would probably be partially (e.g. 50 per cent) filled with liquid in which the coils could be immersed. Niobium–titanium superconductors would probably be restricted to maximum temperatures less than 5 K, but even so the cryostat could be designed to be capable of absorbing up to 0.5 kJ per litre of capacity and could therefore be well enough insulated for several hours' operation. With niobium–tin, or other similar superconductors with high critical temperatures (e.g. around 18 K), the cryostat could be designed for final temperatures of around 10 K (at which the pressure would depend on the proportion of liquid present initially, but could reach 25 atmospheres) and for a heat capacity an order of magnitude greater. Thus it should be possible to provide capacity for a normal day's operation.

There still remains the problem of recovering the helium gas, and replenishing the cryostat with liquid helium, with the greatest possible thermal efficiency and minimum possible risk of contaminating the vehicle's cryogenic systems. Although published information is sparse, significant developments have been made in these areas, and there seems little doubt that practical solutions can be found. Indeed, several large-scale experimental systems have been built and operated, as described in Chapter 6, though it is too early to be able to make a critical comparison.

4.7 Magnetic screening

It may already have occurred to the reader that superconducting coils designed to produce a field of several tenths of a tesla at the track may also produce fields of a similar magnitude elsewhere around and inside the vehicle. These stray fields are undesirable on several counts; apart from the obvious drawback of attracting ferromagnetic objects they may produce unwanted biological effects in humans, and they may also interfere with electrical apparatus (either in the control or communication equipment, or personal devices such as heart pacemakers). Although there is no evidence that intense magnetic fields produce immediately obvious or harmful biological effects there is certainly evidence of living creatures being influenced by magnetic fields. The possibility of long-term or subtle effects cannot therefore be discounted. Thus it is unlikely that any system would be acceptable for passenger service unless the stray fields everywhere within the passenger compartment were below a fairly low limit, such as 0.02 T. With this level of stray field any other effects would probably also be within tolerable bounds, though to be completely sure the limit should be set an order of magnitude lower.

In some designs of vehicle, such as those with overhead suspension [Fig. 4.3(c)] or those with magnets only at the ends of the vehicle and so not directly below the passengers, the stray fields will be so low in the sensitive regions that screening problems are largely avoided. On the other hand, in the technically favoured designs, employing linear synchronous motors with an array of magnets the full length of the vehicle, the question of stray fields requires serious consideration. The graph of Fig. 4.11 shows the variation of field above a single coil; although it falls off rapidly with distance the field in a significant part of the passenger compartment would exceed an acceptable limit.

Three methods of screening are possible, namely (i) a sheet of superconductor between the magnets and the passenger compartment, (ii) a screen of magnetically soft material, e.g., mild steel, separating the passengers from the magnets, or (iii) 'bucking' coils located above the main coils, as shown in Fig. 4.12. Of these the first has not been seriously proposed. The second was studied by the MIT team, who concluded[59] that it would be practical but would necessitate an increase in gross vehicle weight of the order of 15 to 30 per cent. A magnetic screen has also been designed for the proposed Canadian vehicle and was estimated to contribute 6 per cent of the gross weight; however, the screening specification chosen appears to be slightly less

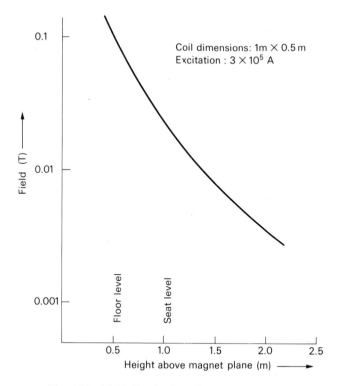

Fig. 4.11   Field distribution above rectangular coil

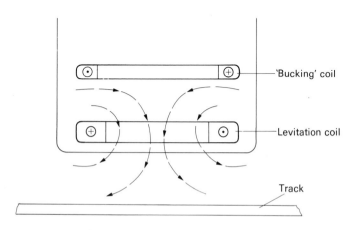

Fig. 4.12   Arrangement of 'Bucking' coil

stringent. The third approach appears to be the most favoured, since good cancellation of the fields in the passenger compartment is possible at the expense of a total increase in the requirement for superconductor of the order of 30 per cent; the resultant weight increase would probably be up to 10 per cent of the vehicle weight.

Thus, although the problem of stray fields can be solved in a practical manner, it will nevertheless be a significant aspect of the design of a revenue vehicle.

5

PROPULSION

5.1 Power requirements

It has already been mentioned in the discussion of guideway design of Chapter 3 that, important though the magnetic drag is, it is still considerably less than the aerodynamic drag on any vehicle moving at a speed of 500 km/h. As the speed is decreased the position reverses, the aerodynamic drag becoming less (being proportional to v^2) while the magnetic drag increases; the resultant drag–speed characteristic is typically of the form shown in Fig. 5.1.

Note that the practical application of Fig. 5.1 is very limited for a number of reasons, amongst which are (i) if a null-flux or hybrid guideway is adopted the magnetic component will be substantially reduced, (ii) the graph assumes constant lift force, whereas in practice in the low speed range the vehicle may be partially supported on wheels, and (iii) under normal operating conditions the vehicle would be running at low speed only over known stretches of track, which could then be designed for low drag by laying down thicker track conductors. Therefore a first discussion of the propulsion requirements can be based on the aerodynamic drag at the maximum operating speed.

The aerodynamic drag is usually normalized, as previously discussed in Section 4.4, such that the force, F_{AD}, is given by:

$$F_{AD} = \tfrac{1}{2} C_D A \rho v^2.$$

As before, A is the frontal area of the vehicle, v the speed, and ρ the density of air (about 1.2 kg/m³). The drag coefficient, C_D, which depends partly on the shape of the vehicle and partly on the surface friction, could theoretically be as low as 0.2 for the type of slender vehicles envisaged for carrying about 100 passengers. However, the need for doors, windows, and various minor protuberances makes it very unlikely that a drag coefficient of less than 0.3 could be achieved in practice. For a train of three vehicles, each of similar capacity, C_D might rise to around 0.6, though the resulting drag would then be a smaller fraction of the overall weight. The curves of Fig. 5.1 have been calculated for a 30 t vehicle using $C_D = 0.3$, $A = 9$ m², and magnetic lift/drag ratio of 20 at a speed of 500 km/h.

It may be concluded that any form of very high-speed transport will require large amounts of propulsive power, in the region of 6 or 7 MW for a 100-passenger vehicle.[60] Provision and transmission of this power present many problems, whatever the mode of transport. For example, one objection sometimes made against the use of steel wheels and rails at very high speeds is

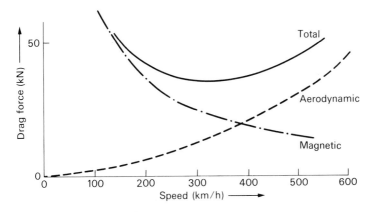

Fig. 5.1   Drag force versus speed

the difficulty of transmitting mechanical traction forces reliably by means of wheel/rail friction.

In the case of magnetically levitated vehicles non-contact forms of propulsion must be considered, and these fall broadly into three classes, namely:

(i)   an on-board prime mover;
(ii) 'short-stator' linear electric motors with the active part on the vehicle, and thus with power collection from sliding contacts overhead or at the track side, and
(iii) 'long-stator' linear motors with a powered track winding and a passive vehicle.

5.2   Powered vehicles

On-board prime movers offer the most straightforward solution for propulsion, especially if used directly as, for example, jet propulsion from gas turbines, airscrews, or ducted fans, as employed on some experimental tracked hovercraft. Alternatively, they may be used indirectly to drive electrical generators to power the on-board stator windings of a linear induction machine. Yet another proposal[61] is the magnetic 'paddle wheel' (using superconducting magnets) suggested by the Ford group, which would also be driven directly by a prime mover.

All such schemes tend to suffer from the disadvantages of noise, pollution from exhaust gases, and the need to incorporate a large weight and volume of both equipment and fuel storage into the vehicle. The last objections are particularly relevant to the use of on-board generators coupled with short-stator linear induction motors, since conventional electrical machines employing iron cores have a very poor power/weight ratio. For example, gas turbines weigh around 1 t/MW, whereas conventional railway drives based on diesel-generators and d.c. motors weigh up to 10 t/MW. Nevertheless, by putting all the power generating and propulsion equipment on board the vehicle the very severe problem of high-speed power collection is avoided and the track is reduced to its most basic and most simple form. This could indeed be the best solution in sparsely populated areas, where the environmental problems are less restrictive.

Against the remarks of the preceding paragraphs it should be recalled that the original surge of interest in magnetic levitation ten years ago arose from the desire to avoid the problems of increasing air traffic and overcrowded air lanes in areas such as the North Eastern United States and Western Europe without sacrificing overall journey speeds. The tracks would thus have to run through densely populated areas and, in this case, electric linear motor propulsion would have definite advantages. The most important of these are that it is relatively clean (any pollution is confined to central power stations), relatively quiet (the noise level arises solely from aerodynamic effects, though for a vehicle moving at 500 km/h it may nevertheless be comparable with the mechanical noise level of much lower-speed conventional vehicles), and flexible in operation in that it is not restricted to one primary fuel. A further advantage is that it requires only a minimum of maintenance, since there are virtually no mechanically moving parts.

Where the power is utilized on board the vehicle, such as for a linear induction drive, there are serious problems of current collection at high speeds through sliding contacts, and these constitute a major drawback to such schemes. Overhead systems using pantographs for power collection are being continuously developed for operation at ever higher speeds, but service speeds of 500 km/h are well above those that can at present be envisaged. An alternative, the type of rigid track 3-phase current distribution and collector shoe used on the US High-Speed Test Track at Pueblo, Colorado, could perhaps more readily be developed. However, as in neither case are there any running rails present to act as a return conductor, at least two, and preferably three, supply lines would be needed, so the overall cost of electrification may be expected to be high. Also, depending on the design of linear machine adopted, a considerable weight of conversion equipment may be needed on-board to provide a variable-voltage and variable-frequency supply to allow for effective operation over a wide speed range. Any increase in weight and volume is of considerable importance, since it brings penalties both in increased magnetic drag and in increased civil engineering costs for guideway construction and maintenance, as well as limiting the payload of the vehicles.

5.3   Passive vehicles

It is possible to avoid some of the problems associated with electrical propulsion discussed in the preceding section, by laying the powered side of the electrical machine on the track so that the coupling to the vehicle is through electromagnetic interaction. This is the approach most favoured in the development of magnetic levitation systems. Within this approach variants on most of the well-known electrical machines are possible. For example, an induction machine may be built with a three-phase winding (preferably iron-cored) along the track, creating a travelling magnetic wave, which interacts with a passive reaction plate on the vehicle; however, for reasons discussed below, there are now no serious developments of this design. Likewise, a synchronous machine may consist of a similar three-phase winding (now iron-free) along the track and a d.c. field winding on the vehicle. In most designs of the latter the field would be provided by superconducting magnets, thus minimizing the on-board power requirement and, as shown below, permitting high efficiency; such machines are receiving considerable attention in a number of countries. Other linear machines proposed, though not at present being developed, are the homopolar[8] (which suffers from the need for sliding contacts), and the d.c. commutator motor with electronic commutation.[62]

In the past, linear induction machines have aroused considerable interest, both as a means of propulsion combined with a separate suspension, and as a combined propulsion, levitation, and guidance system such as the 'Magnetic River' proposed by Laithwaite.[3] Indeed, a number of the former have been built to power high-speed test vehicles. However, even with the powered half of the machine (the 'stator') on the vehicle, and thus relatively short, and with a relatively small air gap (e.g. 10 mm) between the vehicle and the reaction rail (the 'rotor') it is still not easy to design for good efficiency and power factor.[60] The position is many times worse with a powered, i.e., long-stator, track and with large clearances. The reason is of course the very poor magnetic coupling between the primary, or stator, and the secondary reaction plate on the vehicle, so that very large track currents must flow in order to induce enough current in the reaction plate to generate the desired thrust; on top of this the flow of magnetic energy into and out of the air gap results in a large 'reactive power' demand and hence poor power factor. Therefore this configuration of linear induction machine can be dismissed from any further consideration for high-speed transport applications.

The more efficient long-stator linear synchronous machine (LSM), and the commutator machine, which is closely related, tend to suffer from the same problem of poor magnetic coupling outlined above. Now, however, the price can be paid by increasing the excitation of the field coils on the vehicle, which with superconducting coils is relatively easily done. Several designs have been proposed in some detail and in all of these it is possible to supply several kilometres of track winding with current at power frequencies, and still achieve very high efficiencies and power factors. The quantity of cryogenic equipment—superconducting coils and cryostats—required for this is not inconsiderable; typically it will be an order of magnitude greater than that required for levitation by the simplest conducting-sheet track. For example, the Canadian design[37] employs 10 magnets for levitation and 50 for LSM propulsion. On the other hand, it has already been shown that practical systems would probably use more magnets than the simplest arrangement in order to give guidance and perhaps also to reduce the magnetic drag, so the imbalance between propulsion and suspension would be lessened. Indeed, in the MIT and University of Warwick designs the one set of magnets serves all three functions, and in the Canadian design cited the propulsion magnets are also used to generate guidance forces.

5.4   Linear synchronous machines

5.4.1   *Design of LSMs*

To put matters into perspective it is useful to consider specific designs of the linear synchronous machine. One important parameter is the magnetic wavelength, λ (or the corresponding pole pitch $\lambda/2$), since it relates the synchronous speed v to the supply frequency f by $v = \lambda f$. Optimization of the choice of wavelength has been discussed in several papers and reports. For example, by maximizing the expression for the thrust per pole the MIT group found[47] a value of λ of about 3 m, whereas the Canadian group, considering the total thrust on a vehicle of given length (so the number of poles on the vehicle contributing to the thrust becomes inversely proportional to the magnetic wavelength), found[37] an optimum of about 1 m. However, there can be yet other variants on the optimization procedure. If the field at the track is supposed given, instead of depending, albeit rather weakly, on the wavelength, the overall cost is found to vary slowly but monotonically with wavelength and hence there is no optimum. Nevertheless, in all cases the range of acceptable values of λ is wide, and a final choice will depend on many factors outside simple optimizations. Thus, wavelengths in the range 1 to 3 m are to be expected. The magnets on the vehicle are generally alternating in sense, so that individual coil lengths are somewhat less than half a wavelength. Their width is determined by the geometry of the levitation strips since, although the greatest width possible is desired on grounds of efficiency, it is also necessary to avoid interaction, and the consequent power losses, between the stator (track) winding and the levitation strips. As an example, the Canadian design has a stator winding 1.6 m wide situated in the 1.95 m gap between the levitation strips on the guideway, the vehicle coils being of mean width 1.7 m. For an integrated scheme such as the MIT Magneplane or the Warwick University proposals the stator winding would be similar, but the vehicle coils would have to be somewhat wider to overlap the levitation strips (see, for example, Fig. 4.4).

The track armature winding is usually 3-phase, and, in its simplest form [Fig. 5.2(a)], consists of three wave windings each displaced $\lambda/3$ from the other two. In conventional rotary machines there are many conductors per pole connected or wound so that the terminal voltage is the (vector) sum of the individual voltages induced, and so can be matched to the desired supply voltage. In a representative design of linear synchronous machine for maglev application the terminal voltage and current requirements of a single conductor are about 6 kV and 500 A, so multiple conductors in series are quite unnecessary. However, there are some advantages in using more than one conductor—for example, the Canadian design has two conductors per pole separated by about $\frac{1}{6}$ pole pitch [Fig. 5.2(b)]; flexibility of supply conditions is afforded by the choice of either series or parallel connection, and the self-inductance of the winding, which appears as leakage inductance in the equivalent circuit of the machine and thus degrades its behaviour, can be significantly reduced. The detailed design includes such requirements as periodic transposition of the conductors to ensure balance in the parallel connection, and subdivision of each conductor (of about 11 mm diameter)

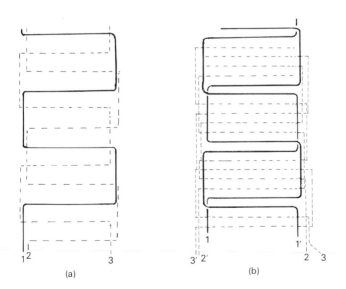

Fig. 5.2   Plan of linear motor track winding
(a) Simple 3-phase wave winding
(b) Two conductors per phase

into many strands (e.g. 37 or 61) to minimize eddy-current effects. It is interesting to note that, in connection with this latter point, experimental evidence has shown that insulation of the individual strands is not necessary. The reduction in self-inductance can also be achieved by using a wide tape conductor, as in the MIT proposals.

A further optimization of the linear motor concerns the spacing of power conditioning equipment along the track side. Detailed studies have been made by the MIT and the Canadian research groups, and more recently by the Germans and the Japanese, of the balance between the capital cost of the individual power units, on the one hand, and the increase in losses and reactive power requirements brought about by an increase in the length of track energized at any one time on the other. The analysis is clearly dependent on factors such as the frequency of traffic, equipment and energy costs, and the degree of reliability (achieved by duplication of equipment) sought. Nevertheless, for realistic values but assuming a fairly high traffic density (an *average* loading of one train every 5 minutes) the optimum energized length is between 2.5 and 5 km. Shorter lengths would give a slightly higher efficiency but, in order to halve the overall losses (from 70 per cent efficiency to 85 per cent), the energized length must be reduced by a factor of 10.

5.4.2 *Control of LSM*

In Section 4.5 it was pointed out that the power equipment, as well as providing the thrust force, can also be used to provide active control of vehicle heave motion, thereby improving the ride quality. Thus the individual power units would comprise step-down transformers, thyristor switches (arranged either as a current-controlled inverter by the Canadians, or with phase-controlled switching of the 60 Hz supply for the majority of the units in the MIT proposals), control equipment, and any necessary power factor correction equipment. The Canadians achieve this by controlling the power angle of the machine (that is, the phase of the travelling magnetic wave on the track relative to that represented by the vehicle magnets alone) to lie in the normally unstable range above 90°.

In the studies of the power conditioning equipment it was found that the total cost per kVA controlled was about a hundred times the list price of power thyristors. It thus seems reasonable that variants of the linear synchronous machine, such as the thyristor commutated d.c. machine, which use many more thyristors should also be considered. It is difficult to assess the potential of such schemes, though the trend of ever-reducing costs of thyristors and of the integrated circuits needed for control can only favour the idea. A preliminary study[48] made at Warwick University yielded overall figures for costs (both capital and running) rather below those for the synchronous machine, so further investigation is in progress. A similar scheme was proposed some years ago for an urban transport scheme in Australia.[62]

5.4.3 *LSM and wheeled vehicles*

Although the subject of this monograph is magnetic levitation it has, nevertheless, been necessary to consider the types of motor which would be suitable for propelling a levitated vehicle. As implied by the space devoted to it, the combination favoured by most research groups and put forward in some detail in a number of design studies is that of electrodynamic levitation with a linear synchronous motor. However, it must be recognized that the building of such a transport system belongs almost certainly to many years into the future. Apart from any outstanding technical problems, there is an enormous economic barrier set by the costs of new track alignments and the capital investment and other interests represented by the existing railways. It is thus appropriate to examine how far the use of linear motors might improve the performance of trains running on conventional steel wheels and rails.

Maximum operational speeds achieved today by, for example, the Advanced Passenger Train of British Rail are in the vicinity of 200 km/h. In part the speed is limited, mainly through considerations of passenger comfort, by the curvature of existing track alignments. It is also limited by the increased vibration, wear, and maintenance problems, which higher speeds bring, and by the need for the greater propulsion and braking forces to be transmitted by means of friction between wheel and rail.

The use of linear motors would obviously remove the limitations set by friction. Moreover, because of this it would remove also the need for heavy locomotives, or for heavy loads on the driven axles of multiple-unit trains, which in turn would reduce wear and damage to the track. With a long-stator linear synchronous motor employing superconducting magnets for excitation further benefit could result. First, the large air gap of such a machine (in contrast to the much smaller gap needed for iron-cored machines such as the linear induction motor) would permit the magnets to be mounted directly on the vehicle, thus leaving unsprung only the wheels needed for support and guidance and, hence, reducing damage to the track yet further. Second, elimination of the need for collection of electric current would permit improved streamlining of the vehicle and, hence, some reduction in power consumption. Finally, it is to be expected that the vehicles would be of lighter construction than conventional railway coaches, with again a reduction in wear of the track.

The speed, acceleration, and braking (through regenerative operation of the LSM) would now be limited only by considerations of track alignment and passenger comfort. On many parts of routes in the UK speeds in excess of 300 km/h should be easily attainable, leading to significant improvements in journey time. Yet, even though this train would be a radical development in railway engineering, it would nevertheless run on existing rails along with trains composed of present-day rolling stock. The difference in the track would be the addition of the armature winding for the LSM, which would be laid between the rails (possibly embedded in concrete).

The conclusion is that a wheeled train powered by a linear synchronous machine is a natural stage in the development of a magnetically levitated railway. It offers in its own right the prospect of considerable improvement to services in the short term, while allowing the establishment of the technologies needed for the longer-term goal of a fully levitated system.

5.5 Miscellaneous propulsion

For completeness there should be included a number of other methods of propulsion, all of which have been put forward in connection with high-speed transport, though none has reached the stage of serious development. No claim is made to mention all conceivable linear electric drives, many of which have only been considered for low speed or other special applications.

The linear homopolar motor was suggested by Polgreen[8] as being conceptually simple, since the field already present for levitation would also serve as the excitation field. However, the usual drawbacks to homopolar machines, namely, the need for sliding electrical contacts and the low generated voltage for each pair of contacts, cannot be avoided and would seem to rule out the machine for high speeds. Even with the use of superconducting magnets the voltage per armature conductor would only be around 100 V, thus requiring sliding contacts rated for around 50 000 A in total.

Another proposal is that of the Ford group for a superconducting 'paddle wheel'.[61] In essence the paddle blades are superconducting magnets, and as the wheel is rotated over the conducting-sheet track forces are generated electrodynamically. Provided the wheel is rotating fast enough, i.e., the peripheral speed exceeds the vehicle speed, so that the magnets instantaneously nearest the track appear to be moving backwards, the usual magnetic drag force then becomes a thrust. Reasonable efficiencies appear to be possible for wheels of practicable size (e.g., 1 m diameter), rotation speed, and magnetic loading. However, the technical details, such as screening the stray (alternating) field, or the cryogenic engineering of the rotor, have not been discussed in any depth. A related Ford proposal, which was thought to offer higher efficiency, was for a superconducting Archimedean screw, i.e., a helix of superconductor carrying a heavy current, and rotating about its axis which is aligned with the direction of motion.

The drive to the paddle wheel or screw may be a gas turbine or an electric motor. The Ford team appears to prefer the former as being overall cheaper and more efficient, as avoiding problems of power collection, and yet having only a small penalty in increased vehicle weight. Moreover, the gas turbine/paddle wheel is claimed to be appreciably more efficient than the gas turbine/ducted fan. In one of their last papers on the paddle-wheel motor, the

Ford group also analysed the lift forces produced, and showed that these could be large enough to make separate levitation magnets unnecessary. The ensuing reductions in low-speed drag led to the suggestion that, over a representative journey with stops at intervals of 100 to 200 km, the total energy demand of the paddle-wheel scheme might be far less than that of any other hitherto considered.

A final aspect of linear motors, which, with the exception of the Ford work just mentioned, has been largely neglected in cryogenic systems of levitation, is that of the normal or lift forces produced. Heave damping by means of control of the linear synchronous machine is, of course, the notable exception. In principle, however, the relative proportions of lift and thrust may be varied over a wide range, and it is not difficult to design machines in which the vertical force equals the vehicle weight and thus renders any other levitation magnets redundant. The so-called 'Magnetic River'[3] proposed by Laithwaite utilizes this principle with the linear induction motor. Similar behaviour could probably be realized through suitable modification of the linear synchronous or linear commutator machine. It is to be expected that the track currents in these levitating motors will be considerably greater than those in motors designed for thrust alone, and therefore their efficiency and power factor will be degraded. The extra losses must, however, be offset against the elimination of the magnetic drag of separate electrodynamic levitation and, indeed, the elimination of the drag peak is one of the more appealing prospects offered by such schemes. A serious objection is, however, that the inherent stability of electrodynamic levitation is lost, and that automatic control of the armature currents then becomes vital. Perhaps for this reason, further developments in this direction are not to be expected.

6

MAGLEV PRESENT AND FUTURE

6.1 Introduction

Since the original investigations in America of the concept of electrodynamic levitation, using superconducting magnets on the vehicle, to replace the steel wheels and rails of conventional railway trains, a number of major developments have taken place in various other countries. The principal centres of this work are in Japan, Germany, Canada, and, to a small extent, the UK, the scale of effort being, to some degree, directly related to the transport and social problems in the country concerned.

For example, in Japan the National Railways (JNR) Shinkansen inter-city services at maximum speeds of 210 km/h, are rapidly becoming saturated, especially between Tokyo and Osaka, and because of this heavy traffic the costs of track maintenance are imposing a grave financial burden on the system. Perhaps more important are the politically sensitive social pressures which are building up in the country against the noise and vibration associated with the Shinkansen system. It is these latter pressures which are leading JNR, and also transport authorities in other countries, not only to seek solutions applicable to the existing trains but also to investigate alternative technologies which would meet the high-speed requirements by sidestepping the problems of the existing systems.

The same motivation for advanced high-speed ground transport is also present to a lesser extent in the advanced Western countries. In all, the most satisfactory solution to the two basic requirements of high speed and low environmental pollution appears to be offered by the combination of magnetic levitation and linear electric motor. The scope of this monograph is restricted to electrodynamic levitation, and therefore other significant developments using electromagnetic attraction will not be discussed. What will be outlined in this chapter are various research and development programmes which have led to the operation of large-scale test vehicles and to serious designs of passenger-carrying vehicles.

6.2 The USA

In view of the fact that so much of the pioneering of both the basic concepts of magnetic levitation and the experimental studies took place in the USA[18–21, 46] it is appropriate to summarize these activities, even though most of these efforts have now ceased. The principal reason for this decline would appear to be that the American road and air networks are widespread and highly developed, and thus the pressure to introduce a more advanced ground transport system has not seemed to be very great. However, while the first ideas on maglev were being put forward by Powell and Danby[19] and by others, independent studies showed that, in certain parts of the country and notably in the so-called North-East corridor linking Boston, New York, and Washington, a strong case could be made for a future high-speed ground transport system, competitive with aircraft in journey time and superior in total passenger capacity.

The US Department of Transportation supported work centred on the Stanford Research Institute in California[20] and the Ford Motor Company in Detroit,[21] while the National Science Foundation funded a programme at MIT. As well as the enormous contribution to the scientific literature two levitated vehicles resulted—one a towed test vehicle at Stanford,[39] and the other, at MIT,[49] a small-scale model (1 m long, weighing 14 kg) incorporating levitation, guidance, LSM propulsion and active damping through the LSM. In addition, several conceptual designs of revenue vehicles were published, though, as these were perhaps more speculative than those in other countries to be described, no details will be given. It is, however, worth noting firstly the contrast between the Ford studies, which favoured an on-board gas turbine/ducted-fan propulsion, and MIT who preferred the long-stator LSM and, secondly, the choice by both groups of guideway cross-sections which were not flat (inverted-T by Ford, and a semicircular trough by MIT).

This work is, unfortunately, now of historical interest, since in early 1975 all the research and development programmes on advanced levitated transport were abruptly stopped by the withdrawal of government funding. Little progress has been reported since that time, though the most recent energy crisis has revived interest in ground transport modes which do not rely on oil or any other specific energy source, and there are indications that the development work may soon be taken up again with renewed vigour.

6.3 Japan

6.3.1 *History*

In terms of scale and capital investment the most advanced development of electrodynamic levitation is perhaps that of the Japanese National Railways. In 1962 studies began on ways in which reliance on the frictional adhesion between wheel and rail might be avoided, and on the use of linear motors for propulsion. In the early stages of the development effort was focused on the short-stator linear induction motor (LIM)—that is, with the powered 'stator' on the vehicle and reaction rail on the track. However, as the work progressed it became clear that the linear synchronous motor (LSM) would be more compatible with the electrodynamic levitation.

Studies continued on the design of lightweight compact superconducting magnets for both the levitation and the propulsion of the vehicle. Several large-scale test facilities, including rotating-wheel rigs and linear test tracks, have been constructed and the work has now culminated in the building of a vehicle (designated ML500) designed to run at 500 km/h on a full-scale test track set up near Miyazaki, Kyushu, in southern Japan.[38] This part of the programme began in 1974, and running tests were started in 1977 on the initial 4.7 km long section. A significant feature of the project is that tests were to be carried out on near actual-size components, at speeds up to the maximum design value. The vehicle and test track are shown in Fig. 6.1, while the general specification of the test facility is summarized in Table 6.1.

6.3.2   The Miyazaki test facility

The streamlined 10 t vehicle has retractable rubber-tyred wheels for suspension and guidance at low speeds and for raising the vehicle to overcome the low-speed magnetic drag peak. Braking is achieved primarily through the linear motor, but this is backed up for safety by a friction brake engaging an I-beam along the top of the central limb of the T-shaped guideway. The vehicle is also provided with skids for emergency landing. Four L-shaped helium

Fig. 6.1   Japanese test vehicle ML-500 on track

cryostats house the superconducting coils. Each cryostat contains two racetrack-type coils mounted horizontally for levitation, and two coils mounted vertically for guidance and propulsion, making a total of 16 magnets on the vehicle. In contrast with developments in other countries, which use a simple aluminium strip track for levitation, the lifting coils on the Miyazaki vehicle react with two rows of discrete coils (wound with aluminium) along the guideway. The coils are prefabricated to close tolerances and accurately positioned on the track.

The operation of the cryostats is interesting, in that for the initial tests the isochoric, or sealed, mode was used. After partially filling with liquid helium at 4.2 K and 1 atmosphere pressure the cryostat was sealed; the pressure and temperature then rose to approximately 2 atmospheres and 5.2 K, respectively, over 2 hours testing. The permissible temperature rise is, of course, set by the characteristics of the niobium–titanium superconductor.

Table 6.1   *Miyazaki test facility*

Track:	
Length	7 km
Levitating coil pitch	0.7 m
Levitating coil dimensions	0.45 × 0.33 m
Guidance/propulsion coil pitch	1.4 m
Guidance/propulsion coil dimensions	1.1 m × 0.7 m
Vehicle:	
Length	13.5 m
Width	3.7 m
Height	2.6 m
Weight	10 t
Speed	500 km/h
Levitation height (coil centre–centre)	250 mm
Total number of levitating coils	8
Levitating coil length	1.65 m
Levitating coil width	0.3 m
Levitating coil pitch	2.1 m
Levitating coil excitation	250 kA-turns
Total number of guidance coils	8
Guidance coil length	1.65 m
Guidance coil width	0.5 m
Guidance coil pitch	2.1 m
Guidance coil excitation	450 kA-turns
Power equipment:	
Motor-generator frequency-changer input	10 MW 60 Hz
Motor-generator frequency-changer output	25 MVA 120 Hz
Cycloconverter input	120 Hz
Cycloconverter output	0–33 Hz

The pressurized helium is then recovered into a wayside storage tank for reliquefaction, and the cryostat recharged with liquid. It is eventually planned to operate the magnets with an on-board refrigerator, as would be proposed for a revenue vehicle, but as a short-term expedient some of the more recent tests have been carried out with the cryostat vented to atmosphere, so the helium simply boiled off and escaped. In this way up to $8\frac{1}{2}$ hours testing was possible before a refill of liquid was required. During the very large number of tests which have so far been carried out, no quenching of the 16 magnets has occurred.

A further feature of the Japanese design which is of considerable interest is the combined guidance and propulsion scheme. For guidance it is quite possible to generate transverse repulsive forces in a similar manner to the levitation, but using a vertically-mounted sheet or series of loops on the guideway. Indeed, the early American proposals were for the L or U-shaped tracks described in Section 4.1. The main disadvantage of these configurations, apart from the large amounts of aluminium required on the track, is the increased magnetic drag force which is created regardless of the net lateral force being generated. A significant improvement, in this respect, is afforded by the null-flux principle of Powell and Danby, and, as has been seen, such schemes for lateral guidance are being developed by the Canadian and the German teams.

The Japanese have likewise adopted the null-flux principle, but have ingeniously incorporated it directly into the armature windings of the linear motor,[38] instead of having separate guidance coils on the track. It is generally accepted that the long-stator linear synchronous motor is the propulsion most compatible with electrodynamic levitation. On the Miyazaki track the motor armature winding is arranged along the track in a double row, one on each side of the central limb, and cross-connected as shown in Fig. 6.2. Normally, with no lateral displacement of the vehicle, the voltages V_1 and V_2 induced in the left and right propulsion coils, respectively, are equal, and the aggregate of such voltages along the track equals the terminal voltage of the motor. If the vehicle shifts sideways the voltages are no longer equal, and their difference gives rise to a circulating current which interacts with the superconducting magnets to produce a restoring force which is roughly proportional to the lateral displacement. Tests have confirmed that there is no interference between this null-flux guidance and the operation of the linear motor propulsion. By using the one set of coils for both functions in this way an economical and reliable system is obtained, for even if the armature power supply fails the guidance will still be maintained.

Unlike the normal practice for a three-phase armature winding there is no overlapping of the phases, the guideway coils being spaced apart. Although this gives rise to significant fluctuations in the thrust forces generated, it has the advantage of simplicity of fabrication and installation of the track coils and

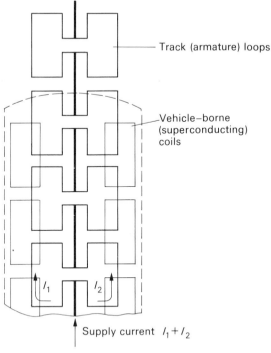

Track (armature) loops

Vehicle–borne (superconducting) coils

Supply current $I_1 + I_2$

Vehicle central: $I_1 = I_2$

Vehicle displaced laterally: I_1, I_2 unequal

Fig. 6.2   Scheme of combined guidance and propulsion

the absence of overlapping of the coils minimizes the winding thickness on the track.

The variable-frequency power supply for the linear motor is provided by a cycloconverter in the substation located halfway along the test line. The highest output frequency is 33 Hz, and the pitch of the coils is arranged for this to produce a maximum cruise speed of 500 km/h. For a revenue system, however, the JNR philosophy is that variable-frequency supplies would be provided only along the accelerating or decelerating parts of the routes, the constant speed cruising sections being supplied at fixed frequency (50 Hz) directly from the grid. By so eliminating the need for power conditioning equipment at frequent intervals the capital costs are considerably reduced. Of course, the problem of restarting the train after an emergency stop remains, but possible solutions are being investigated.

The running of the test vehicle is controlled by a centralized system in the test centre located at the starting end of the track. In addition, facilities for inspecting and repairing the vehicle have been installed, including a helium liquefier with a capacity of 100 l/h. The guideway itself is an elevated structure, the actual runway being of prestressed concrete slabs with reinforcement of high Mn steel which has a low magnetic permeability. A special inspection car is also provided, for ensuring the proper alignment and measurement of the track, power supply systems, and other components.

6.3.3   *Future plans*

Over 2000 test runs have been completed since testing began at Miyazaki, and since the full 7 km length was completed in 1979 speeds of over 500 km/h have been attained. Running at a stable levitation height of between 100 and 120 mm the vehicle is virtually noiseless, and its dynamic stability appears to improve as the speed increases. In general, the test behaviour has been very satisfactory in nearly every respect, all of the major components behaving predictably and effectively.

Hitherto the costs of this programme have been borne by the Japanese National Railways, but now an appreciable part is coming from the Japanese Ministry of Transport and the work is being significantly extended. The inverted T-section guideway has been remodelled into a U-shape (Fig. 6.3) and running tests of a new vehicle, MLU 001-1, the first of a three-car train, commenced by early 1981. There are plans to construct a simulated tunnel entrance for obtaining basic data on the resulting aerodynamic effects on the vehicle, since it is anticipated that the projected route for a revenue system between Tokyo and Osaka would be through tunnels for about 60 per cent of the way. There are also longer-term plans to build a 40 km length of track for testing a full-scale prototype train.

A major area of continuing research and development in Japan is superconducting magnets, cryostats, and helium refrigerators. This is only to be expected, since these relatively novel technologies are crucial to the practical realization of electrodynamic levitation. It is therefore noteworthy that throughout the test programme the magnets functioned reliably, with no signs of quenching during operation of the vehicle. Thermal losses attributable to eddy currents induced by the vehicle vibrations were found to be negligible, and likewise the rate of decay of current in the persistent mode magnets was very small. In the full-scale system it is planned to use on-board refrigeration for the magnets, and it is perhaps here that most development is needed. The alternative of isochoric operation of the cryostats has been shown to work, but for practical operation only the higher temperatures, namely 10 to 12 K, permitted by the use of niobium–tin conductors would allow sufficiently long intervals between helium fills. One result of the Miyazaki programme has been to suggest that the problem of temperature stratification, which might be expected to limit the use of sealed cryostats, is not serious—perhaps because the motion of the vehicle promotes mixing of the helium. A compromise being

rail. However, the existing railway system, on which 90 per cent of traffic is freight, is approaching saturation capacity, so greatly increased passenger services can only be provided by investing in new track alignments. For a dedicated passenger service on a completely new alignment the costs of either a conventional wheel-on-rail system or an advanced, non-contact, maglev system would appear to be similar. This is perhaps not surprising, since the costs are in each case dominated by the civil engineering work involved. The consensus of opinion in Canada is that the benefits of the new technology could greatly exceed those of marginal improvements to current technology, and in the long run may well be the more economical approach. The choice is further narrowed by the severe winter conditions in Canada, for which the electrodynamic levitation with its large clearance between vehicle and track is thought to be more suitable than the narrow-gap alternatives.

The Canadian Maglev Group,[37] a multi-disciplinary group of research scientists and engineers from three universities, Toronto, McGill, and Queen's, working under the aegis of the Canadian Institute for Guided Ground Transportation (CIGGT) at Queen's University, was set up in 1971 by the Transport Canada Research and Development Centre. The establishment of such an organization was thought to offer several benefits, namely (i) ready access to a pool of technical expertise should assist in transport planning, (ii) by promoting interest in innovative concepts, it would both stimulate research into more conventional transport and generate a greater awareness of the needs and challenge of the future, and (iii) such a group would serve as a focus for the interchange of ideas and information with other countries. Indeed, Canada has clearly established herself as an equal partner on the international scene. In the last eight years a considerable amount of authoritative theoretical analysis and laboratory experimentation on nearly all aspects of electrodynamic levitation and linear synchronous motor propulsion has been produced. The focal point of the experimental work has been a large test facility at Queen's University, comprising a stationary superconducting magnet mounted in a force balance and a horizontally mounted rotating wheel 7.6 m in diameter. This was funded by the National Research Council (NRC). Exhaustive tests were carried out on the wheel, over a wide range of operating conditions, and excellent agreement was found between the experimental results and the theoretical predictions of the behaviour of both levitation and linear motor. Also studied were interactions between the magnet and the steel reinforcement needed in structural concrete. Further NRC support has allowed this to lead to the design in some detail of a 100-passenger, 480 km/h revenue vehicle and its major components, as shown in Fig. 6.4 and summarized in Table 6.2. Such vehicles would be run separately on elevated guideways, with the two directions separated to minimize aerodynamic interference. NRC is now playing a continuing role by helping to bridge the gap between university research and industrial development. A particular

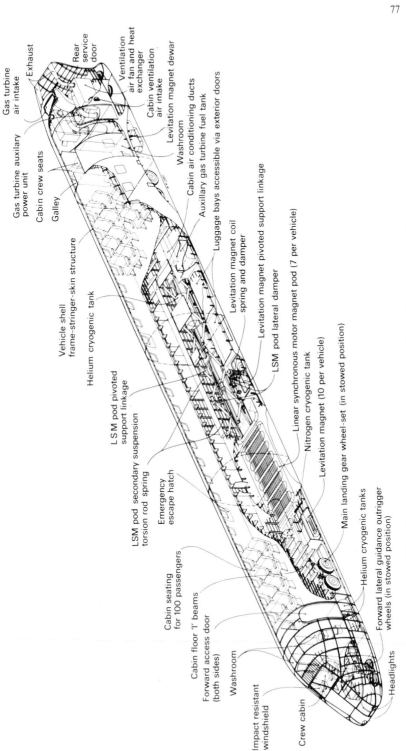

Gas turbine air intake
Exhaust
Rear service door
Ventilation air fan and heat exchanger
Cabin ventilation air intake
Levitation magnet dewar
Washroom
Cabin air conditioning ducts
Auxillary gas turbine fuel tank
Luggage bays accessible via exterior doors
Levitation magnet coil spring and damper
Levitation magnet pivoted support linkage
LSM pod lateral damper
Linear synchronous motor magnet pod (7 per vehicle)
Nitrogen cryogenic tank
Levitation magnet (10 per vehicle)
Main landing gear wheel-set (in stowed position)
Helium cryogenic tanks
Forward lateral guidance outrigger wheels (in stowed position)
Headlights
Crew cabin
Impact resistant windshield
Washroom
Forward access door (both sides)
Cabin floor 'I' beams
Cabin seating for 100 passengers
Emergency escape hatch
LSM pod secondary suspension torsion rod spring
LSM pod pivoted support linkage
Helium cryogenic tank
Vehicle shell frame-stringer-skin structure
Galley
Cabin crew seats
Gas turbine auxilary power unit

Fig. 6.4 Conceptual design of proposed Canadian revenue vehicle

Table 6.2   *Canadian maglev system parameters*

Vehicle:	
Passenger capacity	100
Length	36.5 m
Width × height	3.2 m × 3.2 m
Estimated laden weight	300 kN
Maximum cruising speed	480 km/h
Guideway clearance (above 50 km/h)	15 cm
Estimated aerodynamic drag (at 480 km/h)	29 kN
Electrodynamic suspension:	
Number of superconducting magnets	10
Strength	3.08×10^5 ampere-turns
Size	1.06 m long × 0.30 m wide
Levitation strips	0.6 m wide × 1 cm thick
Magnet suspension height	22 cm
Suspension stiffness	3.0×10^6 N/m
Magnetic drag (at 480 km/h)	13 kN
LSM propulsion:	
Equivalent number of superconducting magnets	50
Strength	5×10^5 ampere-turns
Size	0.53 m long × 1.70 m wide
Pitch	0.57 m
Thrust	42 kN
Guideway stator section length	5 km
Conductor cross-section	1 cm^2
Efficiency	0.75
Power factor	0.93
Guidance:	
Null-flux loop length	0.57 m
Pitch	0.315 m
Conductor cross-section	3.8 cm^2
Estimated 100 km/h side wind load	70 kN
Maximum electrodynamic restoring force	300 kN
Lateral stiffness	2.6×10^6 N/m

problem being investigated in collaboration with industry is the construction of the lightweight magnets which will be needed.

6.4.2   *The proposed vehicle*

The proposed vehicle has separate magnets for levitation and propulsion. Five magnets are distributed along each side of the vehicle to give a clearance of about 150 mm from the guideway. At the cruising speed of 480 km/h the calculated magnetic drag force is 13 kN (lift/drag ratio of 23), while the aerodynamic drag is estimated to be 29 kN. However, in order to counter the speed dependence of the drag and so allow the motor to be designed to produce constant thrust during acceleration, the thickness of the levitation strips on the track is graded from 10 mm in the high-speed cruising sections to 30 mm at lowest speeds. Below 50 km/h, in the very low-speed accelerating regions outside stations, the vehicle will be supported on retractable rubber-tyred wheels, and the aluminium strip omitted in order to avoid the drag peak which is characteristic of this type of levitation.

As with other electrodynamically levitated vehicles, a long-stator air-cored LSM is used for propulsion. The field winding is made up of a series of 50 race-track-wound superconducting coils, arranged in seven pods mounted centrally along the underside of the vehicle. To minimize interaction between pods, one end magnet in each set is made both narrower and weaker. The three-phase armature winding is embedded in the track between the levitation strips; it has two conductors per phase, as illustrated in Fig. 5.2(b). At the cruising speed the propulsion power is about 5.6 MW, while during acceleration up to 70 kN thrust is needed. On-board systems are powered by an auxiliary 75 kW gas turbine mounted at the rear of the vehicle. The variable-frequency power supply is obtained from trackside inverters spaced at 5 km intervals, sections being energized sequentially as the train progresses. The operation of the vehicle—acceleration, cruise, deceleration—is controlled from a terminal computer, so no vehicle–ground link is required.

Because the system is synchronous, vehicle heading and position can be controlled accurately at all times. A further function of the control system is to provide heave damping, and the possibility of allowing damping in other modes, such as roll and yaw, by the use of a double armature winding, has been investigated.

Guidance is by the null-flux scheme described in Section 4.1 and illustrated in Fig. 4.5, in which the propulsion magnets interact with figure-of-eight loops overlying the track winding. However, to minimize force fluctuations there are two layers of loops, overlapped. In the event of very large sideways displacements, such as due to strong side winds, an additional, back-up force is provided by the interaction of the propulsion magnets with the edges of the levitation strips. This is the same type of edge interaction that provides the whole of the guidance in the Warwick split-track design. For low-speed operation, outrigger wheels are provided, engaging steel I-beams at waist height at the side of the guideway. These and side-mounted skids also serve in the event of complete failure of the cryogenic systems to keep the vehicle on the guideway at all speeds.

The lack of inherent damping in the electrodynamic suspension is overcome by a combination of means. Some damping is provided by feedback control of the linear motor armature supply, but to ensure dynamic stability an actively controlled secondary suspension is also considered to be necessary. The alternative would be to construct and maintain the guideway to unacceptably close tolerances. The parameters of the secondary suspension have been optimized in the design to give a ride quality better than the UTACV standard

(Fig. 4.8). The design is realized by mounting each lift magnet on a vertical spring with a parallel hydraulic actuator (Fig. 6.5); normally the actuator is part of an active control loop, though the suspension still functions safely in the event of control system failure. To meet the ride quality specification for the lateral motion the motor magnet pods are mounted on a passive torsion bar and damper suspension (Fig. 6.6), so that they are independently constrained to move only sideways and form the primary suspension mass for this vibration mode.

To avoid any possible weight penalty and low reliability of on-board refrigerators and compressors, the Canadian team advocates isochoric, or sealed, operation of the cryostats on the vehicle. By using niobium–tin conductors for the magnet coils and letting the helium temperature and pressure rise to about 10 K and 15 atmospheres, respectively, a full day's

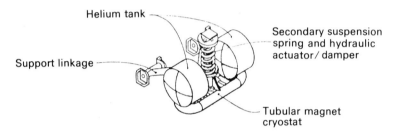

Fig. 6.5   Levitation magnet and support linkage

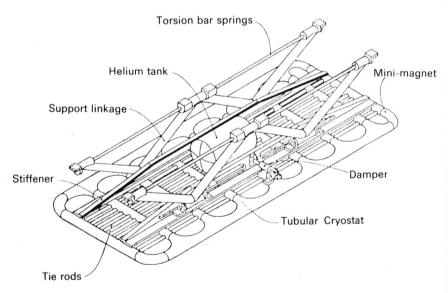

Fig. 6.6   Propulsion magnet support

operation is possible, as the volume of each cryostat is augmented by the adjacent helium reservoir tank.

The superconducting coils would all be operated in the persistent-current mode by means of a mechanical short-circuiting switch inside the cryostat. Recovery of the helium gas, replenishing the liquid helium in the cryostats, and recharging the coils from an external power supply would take place at wayside service stations. In order to contain the pressure the dewar is tubular in section, surrounding the winding, as may be seen from Fig. 6.6. To support the force of repulsion between opposite sides of the race-track coils, fibreglass tie rods are used, located inside the tubes shown and laced at each end with fibreglass into the pancake windings.

To cope with the severe ice and snow conditions which are a regular feature of Canadian winters a flat-topped guideway has been proposed, which avoids the accumulation of snow and debris to which channel or inverted-T sections are prone. Sketched in Fig. 6.7, the structure consists basically of a series of trapezoidal-section concrete box girders on pier supports at 25 m intervals. The top concrete slab is cast *in situ* and the conductors for levitation, guidance, and propulsion embedded in asphalt on the surface. To prevent the generation of eddy currents in the steel reinforcement in the concrete, a network of steel

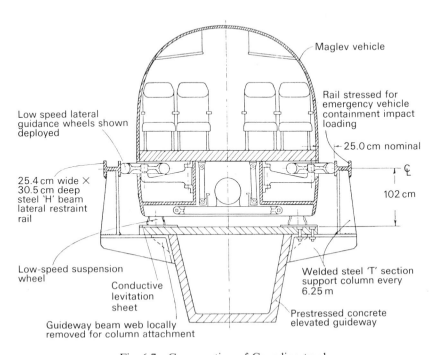

Fig. 6.7   Cross-section of Canadian track

wires is proposed, arranged so that the mesh points are not connected. With the use of standard steel mesh the residual force of attraction between the mesh and the vehicle magnets would be small, while with the use of low-permeability steel it could be negligible. The very small interaction of the magnets with the post-tensioning cable and reinforcement in the lower box section can be ignored.

To provide for stopping vehicles to operate on the same route as high-speed through traffic a switch, or turn-out, has been designed which does not rely on mechanically moving parts. On the guideway in the immediate region of the switch the levitation strips are replaced by discrete coils. Each of these coils, and the null-flux guidance loops, would be capable of being closed by a solid-state switching device. The complete set of guideway conductors for both lines would be superimposed, but only those corresponding to the selected route would be closed, while the others would be left open (Fig. 6.8).

6.4.3   *Future plans*

The basic research programme has been successfully completed and has led to a specific design concept for a maglev transport system to suit Canadian conditions and requirements. The next major stage is the engineering design, prototype construction, and testing of the critical components, of which the superconducting coils are the most notable. In the present phase of the programme a number of different activities are being undertaken. Firstly, a more definitive analysis and assessment of the maglev system performance is being made, including a study of the energy consumption compared with alternative transport systems and optimization studies of the vehicle suspension dynamics and their dependence on guideway tolerances. Secondly,

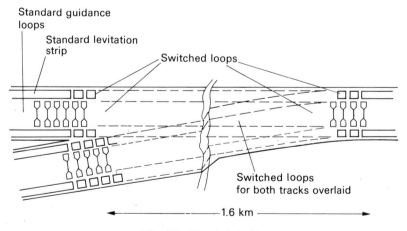

Fig. 6.8   Track switching

the development of superconducting magnets and their associated cryogenic support equipment is proceeding, with emphasis on the operational reliability of a sealed system. Specific items include (i) a flux-pump power supply, incorporating thermally switched Nb_3Sn cryotrons, to compensate current decay in the magnets, (ii) demountable input-current leads, together with a mechanical short-circuiting switch for use in energizing the magnets, (iii) a high-stress test coil to evaluate the conductor and the mechanical support arrangements, and (iv) magnet 'quench' protection schemes suitable for magnets in the persistent current mode. Finally, a detailed economic analysis is being made of the feasibility of the proposed design of revenue system on the Montreal–Ottawa–Toronto corridor. This extends to consideration of a possible time-scale for development and implementation of the system, and identification of the precise changes in circumstances which would make magnetic levitation truly viable for inter-city mass transport.

6.5   Western Germany

6.5.1   *Historical background*

The expansion of the West German economy over the last quarter century has inevitably led to a greater interdependence between the different regions of the country and an increase in the traffic flow. Although the German transport system is one of the most efficient in the world, detailed studies have shown that before the end of the century there will be a gap to be filled between the slowness of road and rail on the one hand and the cost of air transport on the other. A solution to this problem which is economical, and combines high standards of comfort, performance, safety, and reliability with environmental acceptability is precisely what advanced ground transport is intended to provide. As such the problem is, of course, of great interest in most of the advanced countries of the world, though at present the greatest efforts in these new developments are being made in Japan and Germany.

Following the detailed studies of the improvements in traffic between north and south Germany which would result from high-speed ground transport, a major research and development programme[52] was initiated with large funding from the German government. In addition to continuing studies of advanced developments of conventional wheel-on-rail technology, two projects on magnetic levitation are being carried out by separate groups of industrial firms under the sponsorship of the German Federal Government. One of these, Transrapid EMS, funded by the Federal Department of Research and Technology together with the two companies Krauss–Maffei and Messerschmitt–Bolkow–Blohm, is developing the electromagnetic attraction system. The alternative electrodynamic (repulsion) system development is being undertaken by a consortium of the three large electrical companies, viz

AEG–Telefunken, Brown Boveri, and Siemens, in co-operation with the cryogenic company Linde. The Maglev Project Group, comprising specialists from all three firms, was established at the Siemens Research Laboratories at Erlangen in Bavaria. Although both groups have worked on a substantial scale it is only the latter, in which cryogenics plays a decisive role, which is the concern of this monograph.

The principal objectives of the electrodynamic levitation project were the testing of components and subsystems of large-scale dimensions, with the final aim of developing a realistically scaled experimental vehicle. That these aims have been largely realized will be seen from the following.

6.5.2   Erlangen test facility

To be able to undertake extended operation and endurance tests of both the vehicle and track components a large circular track, 280 m in diameter, banked at an angle of 45° to allow for continuous operation at speeds up to 200 km/h, has been constructed at the Siemens site. In cross-section, as shown in Fig. 6.9, the track consists of prefabricated concrete sections forming a C-profile and resting, in turn, on concrete blocks embedded in the ground. The reaction rail for the linear induction motor drive and the aluminium levitation strips are fixed to the concrete by bolts and anchor bars. Some details of the Erlangen test facility are given in Table 6.3. The power supplies, telemetry and data handling and recording systems, together with the helium storage and

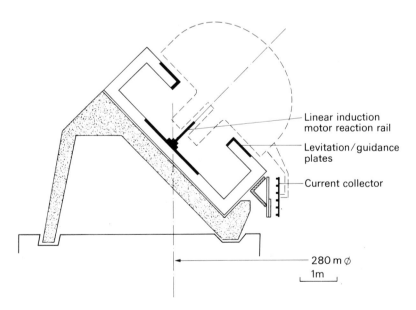

Fig. 6.9   Cross-section of Erlangen test track

transfer equipment, are all located in the control station situated at the track side. Here the unmanned vehicle under test is operated by remote control and data is collected and recorded.

The test vehicle, referred to as EET (Erlangener Erprobungstraeger), has been used in several different ways to test various subsystems. To test drive-systems alone the carrier is supported and guided by a total of sixteen aircraft-type wheels; the same wheels are also used when operating in other modes for landing and starting. The double-sided linear induction motor is flexibly suspended underneath the vehicle, straddling the vertically-mounted aluminium reaction rail positioned along the centre of the track. An accelerating thrust of 45 kN is developed at low speeds, decreasing to about 22 kN at the maximum rated speed of 200 km/h. The a.c. power supply for the carrier is collected from a set of six current-carrying rails mounted along the guideway; the current collector has an air-cushion suspension and is decoupled from the vehicle. The variable-voltage, variable-frequency, power converter for the LIM is also mounted on the test carrier.

The vehicle was designed to carry eight superconducting magnets to provide the lift and guidance forces. These, together with their cryogenic supply systems, are hydraulically mounted on two carrier frames positioned fore and aft on the vehicle. The magnets are racetrack shaped and are operated in the persistent mode by means of a mechanical short-circuiting switch; the excitation is of the order of 500 kA-turns. Details of the magnets are given in

Table 6.3 *Erlangen test facility*

Track	
Diameter, central line	280 m
Banking angle	45°
Reaction plates for lift and guidance:	
material	aluminium
width	600 mm
thickness	20 mm
LIM reaction plates:	
height	800 mm
thickness	11 mm
Test carrier	
Total final mass	18 t
Overall length	12 m
Rated speed	200 km/h
Minimum operational lift gap	100 mm
Gauge	2.6 m
Rated thrust of LIM	22 kN
Converter system	0–105 Hz; 5 MVA
Current rails for LIM supply	3 rails, 3 kV, 1000 A
Current rails, auxiliary	3 rails, 380 V, 200 A

Table 6.4. The magnet cooling system developed by Linde uses liquid forced from a supply tank, pressurized to about 2 atmospheres, to flow through the windings, and then to return through a series of heat exchangers into the refrigerator designed to use the cold fluid as efficiently as possible. The testing of the on-board cryogenic system, involving the complete sequence of cooldown, filling with liquid helium, disconnecting from the stationary supply, two hours operation, and, finally, recoupling to the stationary refrigerator for refilling, has been completed quite successfully. The use of forced cooling for the magnets was thought to be essential because of problems peculiar to the Erlangen test facility, namely (i) the 45° banking angle of the track, which gives rise to different helium levels in adjacent magnets under static conditions, and (ii) the very large centrifugal forces set up at the design speeds which produce liquid helium levels inclined at up to 70° to the horizontal. Forced cooling allows the magnets to be operated in virtually any position, irrespective of such considerations.

Table 6.4   *EET lift magnet: design data*

Conductor:	
Material	Nb Ti 50, Cu (VACRYFLUX)
Cross-section	2.45 mm × 1.4 mm
Critical current (3.5 T/4.2 K)	1000 A
Cu:NbTi	5:1
Twist length	50 mm
Filaments	$300 \times 50 \ \mu m$
Magnet:	
Rated lift force	60 kN
Rated current (3.4 T/5 K)	500 A
Excitation	515 kA-turns
Effective current density	$81 \ A/mm^2$
Maximum flux density	3.4 T
Stored energy	120 kJ
Total mass	540 kg
Mass at 4.3–5 K	200 kg
Winding cross-section	83 mm × 76 mm
Winding length	1.0 m × 0.3 m
Cryostat length	1.4 m × 0.6 m
Height of cryostat body	0.24 m

In the design of the cryogenic system for a full-scale revenue vehicle, however, these same considerations would not necessarily apply, and, hence, other cooling strategies may be envisaged. Of the several possibilities the Erlangen group tend to favour the use of an on-board refrigerator, mainly on grounds of ease of maintenance. A specific design proposal is described below.

A safety feature of the EET is an aluminium ring surrounding each magnet which acts as short-circuited turn. This not only greatly increases the field decay time in the event of an emergency but also helps to reduce losses in the winding by shielding out field variations. Another feature of note is the method of transmitting the mechanical forces across the vacuum space of the cryostat, which is by means of folded (or re-entrant) columns.

The initial testing with the 12 t vehicle, using wheeled suspension without the lift magnets, was begun in 1974, and the design speed of 230 km/h was reached early in 1975. In the second stage of the programme the carrier was equipped with four superconducting lift magnets, and, now weighing 16 t, it was levitated to a height of 100 mm above the track at speeds up to 120 km/h. One disadvantage of circular tracks, highlighted by such tests, is the way the interaction of the centrifugal force and the track banking gives rise to a component of normal reaction which is strongly speed-dependent, leading to loss of levitation beyond an upper speed limit (of around 35 m/s in this case) as well as below the usual low-speed limit.

For the experiments so far mentioned LIM propulsion was used and, apart from some minor problems with the power pick-up, the motor operated very successfully. However, for reasons which have already been discussed in Chapter 5, the decision was taken to investigate the long-stator linear synchronous machine as the preferred propulsion for a revenue vehicle. The resultant design is similar to that of the Canadians, namely an air-cored LSM with an array of superconducting magnets on the vehicle and a three-phase armature winding on the track. In the same way also vehicle guidance is by figure-of-eight null-flux loops on the track, as described in Section 4.3. To verify the theory of the LSM experimentally a large rotating test rig, 5.8 m diameter, built of wood and horizontally mounted, was commissioned. The essential components apart from the wheel are the three-phase armature winding attached to the rim, the superconducting magnet and cryogenic equipment positioned alongside on a mechanical balance (for measuring forces and moments), and the stationary power supply and inverter control equipment. Further details are listed in Table 6.5. Two features of note are (i) that the inductance of each phase of the armature was augmented so as to represent better a full-scale drive, and (ii) two separate armature windings were installed, each fed from its own inverter and occupying half the circumference of the wheel; with this arrangement the transition of a vehicle between powered track sections could be studied. As a result of a comprehensive series of tests no unexpected or untoward behaviour was reported, and the performance was as predicted theoretically to well within the experimental error. In particular, the transition between the two armature sections takes place with no measurable disturbance in either the thrust or lift force up to the highest speeds that were possible.

In view of the success of the rotating rig, larger-scale tests, modelling more closely a full size motor, were planned, and to this end the Erlangen test track has now been converted to allow testing of the long-stator LSM. The vehicle

Table 6.5   *Erlangen rotating test rig (ROSY)*

Diameter	5.8 m
Maximum speed	40 m/s
Stator winding (2 sections):	
Number of turns per phase	4
Pole pitch	0.38 m
Width	0.9 m
Number of pole pitches	24/winding section
PWM inverters (2):	
Maximum power	120 kVA/inverter
Maximum output voltage	390 V
Maximum output current	200 A
Excitation magnet:	
Dimensions	1×0.3 m
Maximum excitation	460 kA-turns
Levitation height	0.13–0.25 m

(Fig. 6.10), now designated EET (02), runs on wheels, and carries two superconducting magnets for the field winding of the LSM. It has an auxiliary 150 kW d.c. drive, which permits open- and short-circuit tests on the synchronous machine and which can act as a load. The magnets are circular coils, of mean diameter 1.2 m, contained in a common outer cryostat; cooling is by liquid helium gravity-fed from a higher-level reservoir which has the capacity for up to 6 hours operation. At the rated current of 1 kA the maximum field is 4.6 T, though the value at the track is much less than this. The 3-phase stator winding is of the usual wave (meander) geometry, consisting of two cables [as Figure 5.2(b)] embedded in concrete supports fixed to the track. This will be clear from Fig. 6.10. The pole pitch is 1.4 m, giving a synchronous speed of 200 km/h at a supply frequency of about 20 Hz. The stator is fed from the pulse-width modulated inverter previously used with the LIM and which is made as two separate units. The units can be connected in parallel to supply the complete armature (860 m) or can be used separately, each supplying only half the armature. The latter arrangement enables the effects of section switching to be investigated. A thrust of 20 kN is developed at the maximum rated current of 825 A at speeds up to 28 m/s. The motor can also be operated in the braking mode, whereby most of the vehicle energy is dissipated in the stator winding, the remainder being absorbed in resistances inserted in the d.c. link of the inverter.

6.5.3   *The next steps*

In addition to the experimental work just described the Erlangen group has also produced[40] a detailed conceptual design of a 200-passenger vehicle, shown in cross-section in Fig. 6.11. The vehicle is made up of two coupled sections, to give an overall length of 56 m and weight of 135 t. Propulsion is

Fig. 6.10   Erlangen test vehicle on track

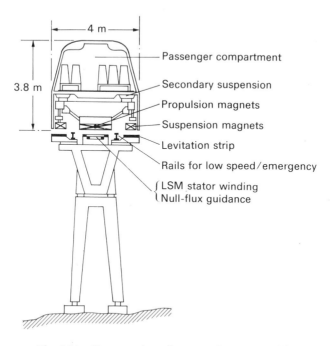

Fig. 6.11   Cross-section of proposed revenue vehicle

provided by 15 magnets in each section, with a design thrust of 87 kN at a speed of 500 km/h; then, on the assumption of a power-factor × efficiency product of 0.66, the total LSM input will be 18 MVA. A track clearance of 100 mm is produced by four pairs of levitation magnets per section of the train. Guidance is obtained by null-flux loops. For low-speed support and guidance there are conventional railway bogies with flanged wheels on steel rails which can be retracted when the train becomes levitated. Although this arrangement has the advantage of compatibility with existing railways it suffers from a considerable weight penalty, and hence increased power demand, as compared with other maglev vehicle designs. The cryogenic system proposed by Linde, sketched in Fig. 6.12, is interesting, in that each pair of levitation magnet cryostats is supplied from one 'cold box' with helium transfer lines made as short as possible, and all four cold boxes on one coach are fed from a single central compressor and control unit. Recently a lightweight rotary compressor has been developed for this application.

As the next stage in the investigation of high-speed magnetic levitation in Germany a new test track, 22 km in length, is being constructed in Emsland in the north of the country.[63] For the initial tests the decision has been taken to proceed with the electromagnetic (attraction) levitation principle. However, the facility is intended for wider use, and the guideway has been designed to enable several types of linear motor to be tested. Thus, in addition to the air-cored LSM described above, the long-stator iron-cored LSM and the short-stator LIM will also be studied.

6.6   The United Kingdom

Since the demise, in 1973, of Tracked Hovercraft Ltd, which had been established for the development of tracked air-cushion vehicles for high-speed

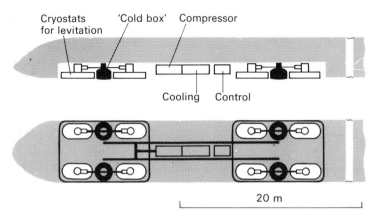

Fig. 6.12   Magnet and refrigerator disposition on revenue vehicle

ground transport, there has been no major government support for this or the more promising electrodynamic levitation. This lack of political will has resulted in maglev research in the UK being on an extremely modest scale compared with that of the preceding sections. The official attitude, as propounded by British Rail, is that the flanged steel wheel-on-steel rail will continue for the foreseeable future, albeit significantly improved through developments such as the High-Speed Train and the Advanced Passenger Train. No doubt the combination of large existing capital investment represented by the present railways and the enormous outlay that would be needed to introduce a new system on any scale have been instrumental in determining this policy.

Despite the lack of official support a research group, funded by the Wolfson Foundation, was established in the early 1970s at Warwick University,[48] for the purpose of investigating the possible application of electrodynamic levitation in high-speed transport. The work has been concentrated on the so-called split-track configuration, which has been described above and for which the potential advantages of obtaining lift, guidance, and propulsion from a single array of magnets on the vehicle interacting with a flat-strip track have been discussed. Although the split-track proposal bears some superficial resemblance to the MIT Magneplane, the lateral stabilizing forces generated by the semicircular guideway of the latter are rather different, in that they appear to arise to a significant extent from the normal reaction between coil and guideway, whereas the Warwick track relies entirely on edge effects.

One of the major difficulties in designing a split track is that, in contrast to the wide flat tracks first proposed for levitation, it is extremely difficult to calculate the eddy-current distribution, and hence the forces that are thereby generated. Therefore much of the initial investigation was undertaken by means of impedance modelling in conjunction with experiments with small models. This was ultimately followed by the construction of the 3 m diameter rotating wheel rig shown in Fig. 6.13. The wheel can be driven or braked, by means of a Ward–Leonard d.c. drive, at rim speeds up to 45 m/s. A typical experimental track consists of two aluminium strips 65 mm wide and 6.35 mm thick, shrunk on to the wheel to leave a central gap 110 mm wide. A superconducting coil, 130 mm square, in its liquid helium cryostat, can be mounted in a force balance located over the wheel. Alternatively, where the problems of measuring the levitation forces and torques (six components in all) in the presence of the dead weight of the cryostat become too great, a liquid nitrogen cooled copper coil can be mounted over the track. The initial results of such tests are encouraging, in that they confirm the behaviour, already suggested as a result of impedance modelling measurements, of a strong dependence of the forces on the levitation height and the gap width, and also show that combinations of track parameters exist which give rise to acceptable combinations of lift, guidance, and drag forces.

Fig. 6.13   Warwick 3 m test wheel

A linear test track 550 m long has also been built, consisting of a timber deck on concrete piers at approximately 5 m intervals (frontispiece). This has been used for aerodynamic tests on streamlined model vehicles typically 3 m long, a major objective being to avoid the problems of imperfect representation of the ground plane inherent in wind-tunnel tests. The models, fitted with suitable instrumentation, are mounted flexibly in a wheeled trolley and towed at speeds up to 50 m/s be means of a continuous cable, driven through a capstan by a conventional petrol engine and gearbox. The track is also provided with a pair of levitation strips, 200 mm wide by 15 mm thick and separated by a gap of 380 mm. The aluminium is in 7 m lengths, clipped on to the timber, with a small expansion gap at the butt joints. To match this track a cryostat containing two square superconducting coils 0.4 m × 0.4 m has been built, but further work on this aspect of the project has been curtailed because of lack of funds.

Small though the Warwick University project is, it nevertheless forms the greatest part of the UK effort on electrodynamic levitation. Other groups have produced one or two paper studies for British Rail or other government bodies, and theoretical investigations into the mixed-μ method of levitation described in Chapter 2 are being carried out, but further effort on any significant scale would seem to be out of the question in the present economic climate.

7

FUTURE PROSPECTS

In this monograph a review of the levitation of moving vehicles by means of electromagnetic forces has been attempted. Research groups, working independently in a number of different countries, have shown how such vehicles, and mass transport systems incorporating them, may be designed for ground speeds up to 500 km/h, thus providing the optimum combination of convenience, fuel efficiency, and speed for the medium-range journeys typical of inter-city travel. Relatively novel technologies, involving superconductors and deep low-temperature refrigeration, are essential for their practical realization, but there appears to be little reason to suppose that this cryogenic technology is incapable of meeting the rather modest demands made on it by maglev.

However, more than technical merit is needed for magnetic levitation to be adopted on any significant scale. In the past, new transport developments have usually offered such overwhelming advantages over existing modes that the entrenched interests of the latter could not long withstand them. Thus, in the last century, the railways in Great Britain developed rapidly at the expense of canals and turnpike roads, once the legal battles to obtain rights-of-way had been won during the two decades up to 1850. Likewise, from about the turn of the century the automobile has supplanted the horse-drawn carriage or cart, following the repeal of the 'Red Flag' Acts which restricted its natural development, and, more recently, it has been a major contributory cause of the decline of the railways. For sea transport the steamship has replaced sail, only to be overtaken, in its turn, by aircraft, at least for passenger transport.

Whether maglev would offer the same degree of advantage as the steam railway did for the Victorian era or the motor car for the present age is something the reader should perhaps judge for himself. What is certainly different now is the general pessimism about the future and an awareness that economic expansion, as we have known it, is rapidly slowing down. Moreover, to be able to prove maglev it is necessary, sooner or later, to work on a significantly larger scale, since the technology is inherently incapable of showing advantage when applied to short routes operated at relatively low speed. The conclusion is that a major political decision, probably at EEC level so far as Europe is concerned, and perhaps even at a wider international level, will be needed. However, although the rate of development of maglev has been reduced in some countries, presumably for economic reasons, large-scale testing is still being continued in Japan and West Germany.

So far as the future prospects for ground transport are concerned, a resurgence of the railway or similar mass transport system would seem to be inevitable; it offers the most satisfactory way of reducing our dependence on oil, while still allowing the demand for travel, which may be expected to continue to increase, to be met. The advantages of an electrified system, which is thus not dependent on any particular primary fuel, have been discussed already in Chapter 1. Also discussed at the same time were the benefits to be had from higher speeds and from the avoidance of mechanical contact between vehicle and track. However these arguments are presented, the inevitable conclusion is that there will be a need for a system of advanced ground transport capable of both satisfying the demand for shorter journey times and being acceptable on environmental grounds. A maglev system, based on one or other of the technologies described in this monograph, would meet these conditions. At present one can be confident about the feasibility of such a solution and give probable costs, though much large-scale development still remains to be done.

However, the most realistic view is that a change to a completely levitated system is far into the future and a more likely course of events is a gradual transition, through linear motors replacing present-day drive mechanisms which rely on wheel–rail friction for traction and braking. For example, with the long-stator linear synchronous machine, discussed in Section 5.4, applied to conventional railways, many of the advantages of maglev could be realized, at least in part. Heavy locomotives, with overhead power collection, would no longer be required. Track maintenance problems, noise, and vibration should be greatly reduced. Significantly higher speeds, better ability to cope with gradients, and central control of vehicle acceleration, braking, and position would all be possible. An ideal application, in which interest has recently been revived, is the proposed Channel Tunnel to link England and France by a single- or double-track railway. Here the gradients at either end would be steep, to keep the tunnel as short as possible. One mode of operation envisaged is a shuttle service (as opposed to running through trains from the normal railway networks of the two countries), in which the acceleration and the possibilities of central control of the LSM could be exploited to advantage. On all counts installation of a linear synchronous drive would seem to lead to more effective use of the track and tunnel.

In conclusion, since the technology for designing a full-scale wheeled train, driven by linear motors, exists at the present time, this would seem to be the best solution, in the short term, to the problem of raising the performance of our railways closer to the standards desired by the travelling public. In the longer term the completely levitated 'railway', capable of speeds up to 500 km/h yet with minimum impact on the environment, promises the ideal solution.

Bibliography

This bibliography is not intended to be complete but, rather, to offer a representative guide to the scientific literature on magnetic levitation.

1. *High Speed Guided Ground Transport of Passengers.* (Parts 1 and 2.) British Rail Technical Report TREDYN 10 (November 1977).
2. The Advanced Passenger Train (APT). *Modern Railways* **34**, (341), 56 (1977).
3. LAITHWAITE, E. R. Three-dimensional engineering. Chapter 11 of *Transport without Wheels.* (ed. LAITHWAITE). Elek, London (1977).
4. IWASA, Y. High speed magnetically levitated & propelled mass ground transportation. Chapter 6 of *Superconducting Machines & Devices.* (ed. FONER, S. and SCHWARTZ, B. B.). Plenum, New York (1974).
5. BLISS, D. S. The evolution of TACVs. *Hovering Craft & Hydrofoil* **9**, (11–12) (1970).
6. GEARY, P. J. *Magnetic and Electric Suspensions.* Taylor & Francis, London (1964).
7. EARNSHAW, S. On the nature of the molecular forces which regulate the constitution of the luminiferous ether. *Trans. Camb. Phil. Soc.* **7**, 97 (1842).
8. POLGREEN, G. R. *New Applications of Modern Magnets.* Macdonald, London (1966).
9. MEISENHOLDER, S. G. and WANG, T. C. *Dynamic Analysis of an Electromagnetic Suspension System for a Suspended Vehicle System.* US Dept. of Transportation Report FRA-RT-73-1 (1972).
10. POLGREEN, G. R. Controlled permanent magnets for tracked transport. *Elect. Rev.* **191**, (2), 41 (1972).
11. SINHA, P. K. and JAYAWANT, B. V. Electromagnetic wheels. *Electronics & Power* 723 (October 1979).
12. LINDER, D. Design & testing of low-speed magnetically suspended vehicles. IEE Conf. Pub. 142, 96 (1976).
    GOODALL, R. M. Suspension and guidance control system for a d.c. attraction maglev vehicle. ibid. 100.
13. GOTTZEIN, E. and CRAMER, W. Critical evaluation of multivariable control techniques based on maglev vehicle design. 4th Symposium on Multivariable Technical Systems, Fredericton, NB (1977).
14. WEH, H., MOSEBACH, H. and MAY, H. Design and technology of the iron-cored LSM for advanced ground transportation. Int. Conf. Elect. Machines, Brussels (1978).
15. Foucault and eddy currents put to service, *The Engineer* **114**, 420 (October 1912).
16. HOCHHÄUSLER, P. The magnetic railway. *ETZ B* **23**, (13), 311 (1971).
17. BRAUNBECK, W. Free vibration of bodies in electrical and magnetic fields. *Z. Phys.* **112**, (11–12), 753 (1939).
18. GUDERJAHN, C. A., WIPF, S. L., FINK, H. J., BOOM, R. W., McKENZIE, K. E., WILLIAMS, D., and DOWNEY, T. Magnetic suspension and guidance for high-speed rockets by superconducting magnets. *J. Appl. Phys.* **40**, (5), 2133 (1969).
19. POWELL, J. R. and DANBY, G. T. A 300 mph magnetically suspended train. *Mech. Eng.* **89**, (11), 30 (1967): or ASME Paper 66-WA/RR-5 (1966). Fig. 2.6 is taken from Chapter 2 of *Superconducting Machines & Devices.* (ed. FONER, S. and SCHWARTZ, B. B.). Plenum, New York (1974).
20. COFFEY, H. T., CHILTON, F. and HOPPIE, L. O. *The Feasibility of Magnetically Levitating High Speed Ground Vehicle.* US Dept. of Transportation Report FRA-RT-72-39 (1972) (NTIS Reference PB210505).

21. DAVIS, L. C., REITZ, J. R., WILKIE, D. F. and BORCHERTS, R. H. *Technical Feasibility of Magnetic Levitation as a Suspension System for High Speed Ground Transportation Vehicles.* US Dept. of Transportation Report FRA-RT-72-40 (1972) (NTIS Reference PB210506).
22. HOMER, G. J., RANDLE, T. C., WALTERS, C. R., WILSON, M. N., and BEVIR, M. K. A new method for stable levitation of an iron body using superconductors. *J. Phys. D: Appl. Phys.* **10**, 879 (1977).
23. SCHOENBERG, D. *Superconductivity.* Cambridge University Press (1965).
24. REITZ, J. R. Forces on moving magnets due to eddy currents. *J. Appl. Phys.* **41**, (5), 2067 (1970).
25. REITZ, J. R. and DAVIS, L. D. Force on a rectangular coil moving above a conducting slab. *J. Appl. Phys.* **43**, (4), 1547 (1972).
26. YAMADA, T. and IWAMOTO, M. Theoretical analysis of lift and drag forces on magnetically suspended high speed trains. *Elect. Eng. in Japan* **92**, (1), 53 (1972).
27. MIERICKE, J. and URANKAR, L. Theory of electrodynamic levitation with a continuous sheet track. *Appl. Phys.* **2**, 201 (1973) and ibid. **3**, 67 (1974).
28. V. HANNAKEM, L. Eddy currents in thin conducting sheets resulting from moving current carrying conductors. *ETZ A* **86**, (13), 427 (1965).
29. RICHARDS, P. L. and TINKHAM, M. Magnetic suspension and propulsion systems for high speed transportation. *J. Appl. Phys.* **43**, (6), 2680 (1972).
30. OOI, B-T. A dynamic circuit theory of the repulsive magnetic levitation system. *Trans. IEEE* **PAS-94**, (3), 994 (1975): or *Trans. IEEE* **MAG-11**, (5), 1495 (1975).
31. DAVIS, L. C. Drag force on a magnet moving near a thin conductor. *J. Appl. Phys.* **43**, (10), 4256 (1972).
32. POWELL, J. R. and DANBY, G. T. Magnetically suspended trains: the application of superconductors to high speed transport. *Cryogenics & Industrial Gases* **4**, (10), 19 (1969).
33. THORNTON, R. D. Design principles for magnetic levitation. *Proc. IEEE* **61**, (5), 586 (1973).
34. DANBY, G. T., JACKSON, J. W., and POWELL, J. R. Force calculations for hybrid (ferro-null flux) low-drag systems. *Trans. IEEE* **MAG-10**, (3), 443 (1974).
35. BORCHERTS, R. H. and DAVIS, L. C. Force on a coil moving over a conducting surface including edge and channel effects. *J. Appl. Phys.* **43**, (5), 2418 (1972).
36. HIERONYMUS, H., MIERICKE, J., PAWLITSCHEK, F., and RUDEL, M. Experimental studies of magnetic forces and null-flux coil arrangements in the inductive levitation system. *Appl. Phys.* **3**, 359 (1974).
37. EASTHAM, A. R. (ed.). Canadian Institute for Guided Ground Transport Report 77-13 (1977): or ATHERTON, D. L., BELANGER, P. R., BURKE, P. E., DAWSON, G. E., EASTHAM, A. R., HAYES, W. F., OOI, B. T., SILVESTER, P. E., and SLEMON, G. R. The Canadian high-speed magnetically levitated vehicle system. *Can. El. Eng. J.* **3**, (2), 3 (1978).
38. KYOTANI, Y. The current state of development of non-contacting suspension and propulsion systems in Japan. Int. Seminar on Superconductive Magnetic Levitated Train, Miyazaki, Japan (November 1978).
39. COFFEY, H. T., COLTON, J. D., and MAHRER, K. D. *Study of a Magnetically Levitated Vehicle.* US Dept. of Transportation Report FRA-RT-73-24 (1973) (NTIS Reference PB221696).
40. ALBRECHT, C. Vehicle levitation and propulsion with superconducting magnets. Concept and State of the West German Project EET. Invited paper given at the 14th Int. Congress. of Refrigeration, Moscow (1975).
41. CAMPBELL, P. and JOHNSON, R. B. I. Forces on magnetically levitated vehicles above flat guideways. IEE Conf. Pub. 142, 121 (1976).

42. WONG, J. Y., MULHALL, B. E., and RHODES, R. G. The impedance modelling technique for investigating the characteristics of electrodynamic levitation systems. *J. Phys. D: Appl. Phys.* **8**, 1948 (1975).
43. MOON, F. C. Laboratory studies of magnetic levitation in the thin track limit. *Trans. IEEE* **MAG-10**, (3), 439 (1974).
44. GOEMANS, P. A. F. M., HAMER, L. K., RAKELS, J. H., SCHOT, J. A. (1978). *Electric Currents in magnetic levitation systems.* Proc. Int. Conf. Electrical Machines, Brussels (September 1978).
45. OHNO, E., IWAMOTO, M., and YAMADA, T. Characteristics of super-conductive magnetic suspension and propulsion for high speed trains. *Proc. IEEE* **61**, (5), 579 (1973).
46. KOLM, H. H. and THORNTON, R. D. Magneplane: guided electromagnetic flight. Proc. Applied Superconductivity Conf., Annapolis (1972 IEEE Cat. No. 72-CHO 682-5TABSC).
47. IWASA, Y., HOENIG, M. O., and KOLM, H. H. Design of a full-scale magneplane vehicle. *Trans. IEEE* **MAG-10**, (3), 402 (1974).
48. RHODES, R. G., MULHALL, B. E., HOWELL, J. P., and ABEL, E. The Wolfson maglev project. *Trans. IEEE* **MAG-10**, (3), 398 (1974): and also *Magnetic Levitation Project Report.* University of Warwick (1976).
49. IWASA, Y., BROWN, W. S., and WALLACE, C. R. An operational 1/25-scale magneplane system with superconducting coils. *Trans. IEEE* **MAG-11**, (5), 1490 (1975).
50. WONG, J. Y., HOWELL, J. P., RHODES, R. G., and MULHALL, B. E. Performance and stability characteristics of an electrodynamically levitated vehicle over a split guideway. *J. Dyn. Syst. Meas. & Control* **98**, 1 (1976).
51. ATHERTON, D. L. and EASTHAM, A. R. Guidance of a high-speed vehicle with electrodynamic suspension. *Trans. IEEE* **MAG-10**, (3), 413 (1974).
52. BOGNER, G. Large-scale applications of superconductivity. Chapter 20 of *Superconducting Applications: SQUIDS and Machines* (ed. SCHWARTZ, B. B. and FONER, S.). Plenum, New York (1977).
53. COFFEY, H. T., SOLINSKY, J. C., COLTON, J. D., and WOODBURY, J. R. Dynamic performance of the SRI maglev vehicle. *Trans. IEEE* **MAG-10**, (5). 451 (1974).
54. SLEMON, G. R. The Canadian maglev project on high-speed inter-urban transport. *Trans. IEEE* **MAG-11**, (5), 1478 (1975).
55. *Performance Specification and Engineering Design Requirements for Urban Tracked Air Cushion Vehicles.* US Dept. of Transportation (May 1971).
56. BROWN, W. S. The effect of long magnets on inductive maglev ride quality. *Trans. IEEE* **MAG-11**, (5), 1498 (1975).
57. HUNT, T. K. A.C. losses in superconducting magnets at low excitation levels. *J. Appl. Phys.* **45**, (2). 907 (1974).
58. MULHALL, B. E. and RHODES, R. G. Sealed liquid helium cryostats for mobile superconducting magnets. *Cryogenics* **16**, (11), 682 (1976).
59. IWASA, Y. Shielding for magnetically levitated vehicles. *Proc. IEEE* **61**, (5), 598 (1973).
60. ABEL, E., MAHTANI, J. L., and RHODES, R. G. Linear machine power requirements and system comparisons. *Trans. IEEE* **MAG-14**, (5), 918 (1978).
61. BORCHERTS, R. H. and DAVIS, L. C. The superconducting paddlewheel as an integrated propulsion levitation machine for high speed ground transportation. *Elect. Machines & Electromechanics* **3**, 341 (1979).
62. LAMB, C. St. J. Analysis and testing of a direct-voltage induced-e.m.f. commutated thyristor motor. *Proc. IEE* **117**, (10), 1975 (1970).

INDEX

A.C. losses, *see* superconductors
AEG, 20, 84
Advanced Passenger Train (APT), **5, 65,** 91
aerodynamic damping, 58
  drag coefficient, 58
  drag force, 18, 30, **58,** 78
  side and rolling forces, 47
Aerotrain, 6
air cushion suspension, 6
armature, *see* linear motor
auxiliary supplies, 54, 79
  suspension, *see* Suspension

Bachelet, 15, 16, 18
Bertin, 6
biological effects of magnetic fields, 55
Braunbeck, 17, 21
British Rail, *see* railways
Brown Boveri Co., 20, 84
'bucking' coil, 55

Canadian Institute for Guided Ground
  Transport (CIGGT), 20, 76
Canadian Maglev Group, 32, 37, 41, 45, 48, 55,
  62–64, 76–83
channel, *see* guideway
  tunnel, 95
comfort, *see* passenger
commutator, *see* linear machine
critical speed, 26
cross wind, *see* wind
cryostat, general details, 23, **53,** 81
  sealed, **54,** 71, 74, 80
current collection, **60,** 85

D.C. motor, *see* linear machine
damping, *see also* aerodynamic, linear motor
  electrical, active, 49, **52,** 79
  electrical, passive, 49, **51**
  mechanical, 49, **50,** 80
Danby, *see* Powell
drag, *see* aerodynamic, electrodynamic
drag peak, 26, 28
  reduction of, **29,** 67, 79

Earnshaw, 9, 17, 21
eddy currents, calculation, 26, 41
  distribution, 24, 36, 43

induction mechanisms, 14, 15, 16
  strength/density, 36
  unwanted, 13, 21, 81
edge effects in guideways, **32,** 39, 46, 91
Eindhoven, Technical University of, 36
electrodynamic forces, calculation of, 24, 26
  drag, 17, 21, **26, 28,** 38, 42, 43, 46
  experimental, **32**
  lateral (guidance), 29, **33, 38, 44**
  lift, 8, 24, 43, 44
  pulsating, in loop track, 37
  roll moment, 46
  speed dependence, 26, 28, 29, 33

ferrite, 9
ferromagnetic hybrid, *see* track
Ford Motor Co., 20, 24, 26, 28, 32, 42, 44, 52,
  59, 66, 69
Fourier methods, 26

guidance force, *see* electrodynamic
  null-flux, 41, **45,** 72, 87
guideway, channel (U section), 7, 39, 45, 74
  costs, 3
  designs, 37*ff*
  inverted-T, 7, 39, 70
  split-track, 39, 46, 91

helium, 16, 53, **54,** 72
Hochhausler, 15
homopolar, *see* linear motor
hovercraft, 6
Hunt, 53

image magnet (or coil), 17, **24**
  force, **24,** 28

Japanese Railways, *see* railways
  and magnetic levitation, 20, 45, 64, 69, 73
joints, *see* track
journey time, 2, 3, 6

Krauss-Maffei, 12, 83

ladder track, *see* track
Laithwaite, 15, 61, 66
lift/drag ratio, 27, 28, 29, 34
  in loop track, 37

linear motor, armature, **63**, 72, 87
  commutator (d.c.), 41, 61, **64,** 67
  homopolar, **10,** 61, **66**
  induction, 5, 8, 14, **15**, 59, **61,** 67, 85
  synchronous, 5, 14, 41, 61, **62***ff,* 79, 87
  synchronous in railways, 65
  synchronous, damping in, **48, 64,** 79
Linde, 20, 84, 86, 90
losses, a.c., *see* superconductor
low speed, *see* suspension *or* drag peak

Magneplane, 20, 39, 41, 42, 62
magnetic field, *see* stray
  keel, 41
  pressure, 16
  river, 14, **15**, 61, 66
  wavelength, 26, 43, 62
mechanical damping, *see* damping
Meissner effect, 22
Messerschmitt–Bolkow–Blohm (MBB), 12, 13, 83
MIT, 20, 31, 38, 39, 41, 42, 45, 48, 55, 62, 64, 69
mixed–μ, 8, **21,** 93
modelling, *see* impedance
  scaling in, 36

National Research Council (NRC) of Canada, 76
niobium–tin, Nb_3Sn, 54, 75, 80
niobium-titanium, NbTi, 54, 71, 86
noise, aerodynamic, 60
  wheel-on-rail, 5
null-flux drag, 31, 33
  guidance, *see* guidance
  levitation, 18, 29
  suspension stiffness, 31, 33
  track design, 30, 31

paddle wheel superconducting motor, 59, 66
passenger comfort, general, 13, 21, **47**
  UTACV standard, **49,** 79
passive damping, *see* damping
permanent magnet, 8, 9, 35, 36, 42
  controlled suspension, 12
pole pitch, 62
Polgreen, 9, 10, 66
Powell and Danby, 18, 30, 31, 69, 72
power equipment, trackside, **64,** 73, 79
power factor, 15, **61,** 62, 67
power spectral density, 49
prime mover, 59
propulsion, power requirement, **58,** 63, 79

Queen's University, 20, 76

railways, British Rail, 5, 12, 65, 91
  Japanese National Railways, 5, 68
  limitations of, 5
  speed of conventional, 4, 5
rare earth–cobalt alloys, 10, 35
refrigeration, **54,** 75, 90
Reitz, 24, 27
revenue vehicle design, Canada, 76*ff*
  Germany, 88
  USA, 69
Richards and Tinkham, 28, 31
ride quality, 13, 21, **48**
Rohr Industries, 15
route switching, 37, 41, **82**

Sandia Laboratories, 18, 20
secondary suspension, *see* suspension
Shinkansen, 4, 68
Siemens AG, 20, 31, 32, 33, 45, 52, 84
skin depth, 17, 24, **26**
  effect, 64
stability, dynamic, 48
  electromagnetic attraction, 9, 10
  quasi-static, 39
Stanford Research Institute (SRI), 20, 32, 45, 52, 53, 69
stray magnetic field, **55,** 66, 82
superconductor, 16, 22, **52,** 54, 55
  a.c. losses, **53,** 87
superconducting magnets, 16, **52,** 71, 85
  stresses in, 53
suspension, auxiliary for low speed, 20
  natural frequency, 44
  secondary, 13, 21, **48***ff*
  stiffness, 20, **31,** 44
Sussex, University of, 12
switching, *see* route *or* thyristor

test centre, Emsland, 90
  Erlangen, 45, 84*ff*
  Myazaki, 70*ff*
  Pueblo, 6, 60
test rigs, 32, 33, 76, 87, 91
Thornton, 31
thyristor, 64
Tinkham, *see* Richards
Tokaido, 4
track, armature winding, 5
  debris on, 21, 81
  hybrid ferromagnetic, 27, 31
  joints in, 43, 93
  ladder, 37
  loop, 37, 71
  thickness, criteria for, 27, 28

travel time, *see* journey
Tracked Hovercraft Limited, 7, 90

Warwick, University of, 41, 45, 46, 47, 62, 64, 79, 91*ff*

wavelength, magnetic, *see* magnetic
wind, effect of cross-, 47, 48
Wolfson, 91